The Changing Dynamics of Energy in the Middle East

The Changing Dynamics of Energy in the Middle East

Volume 1

**Anthony H. Cordesman
and Khalid R. Al-Rodhan**

Published in cooperation with the Center for Strategic and International Studies, Washington, D.C.

**Praeger Security International
Westport, Connecticut · London**

Library of Congress Cataloging-in-Publication Data

Cordesman, Anthony H.
   The changing dynamics of energy in the Middle East / Anthony H. Cordesman and Khalid R. Al-Rodhan.
      p. cm.
   Includes bibliographical references.
   ISBN 0–275–99188–1 (set : alk. paper) — ISBN 0-275-99261-6 (v. 1 : alk. paper) — ISBN 0–275–99262–4 (v. 2 : alk. paper)   1. Petroleum industry and trade—Government policy—Middle East. 2. Petroleum industry and trade—Management. 3. Petroleum products—Prices—Middle East. 4. Petroleum industry and trade Government policy—Africa, North. 5. Petroleum products—Prices—Africa, North.   I. Al-Rodhan, Khalid R. II. Title.
   HD9576.M52C675 2006
   338.2'72820956—dc22   2006022873

British Library Cataloguing in Publication Data is available.

Copyright © 2006 by Center for Strategic and International Studies

All rights reserved. No portion of this book may be reproduced, by any process or technique, without the express written consent of the publisher.

Library of Congress Catalog Card Number: 2006022873
ISBN: 0–275–99188–1 (set)
ISBN: 0–275–99261–6 (vol. I)
ISBN: 0–275–99262–4 (vol. II)

First published in 2006

Praeger Security International, 88 Post Road West, Westport, CT 06881
An imprint of Greenwood Publishing Group, Inc.
www.praeger.com

Printed in the United States of America

The paper used in this book complies with the Permanent Paper Standard issued by the National Information Standards Organization (Z39.48–1984).

10 9 8 7 6 5 4 3 2 1

Contents

Figures

Acknowledgments

This book would have been impossible without the research assistance and work of Emily Fall. The authors would also like to thank Nawaf Obaid for reviewing the text of this book and for his invaluable comments and suggestions on the Saudi section. The authors are grateful for their contribution.

The analysis in this book relied on many sources, including reports by the Energy Information Administration (EIA), the International Energy Agency (IEA), OPEC (Organization of the Petroleum Exporting Countries), United States Geological Survey (USGS), Aramco, British Petroleum (BP), and the analysis of other energy analysts. Data for oil supply and demand were adapted from reports such as the EIA's *International Energy Outlook 2005* and the IEA's *World Energy Outlook 2005.* Reserves data were adapted from three main sources: the BP's *Statistical Review of World Energy 2005,* the IEA's *World Energy Outlook 2002,* and the *U.S. Geological Survey 2000.*

Data provided by the UN Department of Economic and Social Affairs, the World Bank, the International Monetary Fund (IMF), the Saudi American Bank (SAMBA), and other regional agencies on the economic, social, and demographic trends were also of great value in writing this book. Chapter 2 relied on the data provided by the International Institute of Strategic Studies (IISS) *Military Balance,* and the online version of the *CIA World Fact Book 2005.* While the analysis relied heavily on the work of the agencies outlined above, press reports and news articles were used extensively in researching this book.

1

The Importance of MENA Energy

Middle Eastern and North African (MENA) energy exports have become steadily more critical to the global economy and international security. Their importance has risen in recent years as rising Asian demand and a healthy global economy increase demand for oil exports near the limits of global production capacity. Current demand consumes virtually all of the oil that MENA states can produce, in spite of major price increases.

There may well be periods in the future where economic conditions worsen and energy demand and prices drop. The history of energy economics is highly cyclical and filled with periods of boom and bust. Over time, however, it seems likely that growing demand will continue to approach the limits of oil and gas export capacity and keep average prices high. China's and India's growing thirst for oil will be key factors in shaping such demand. However, virtually all energy forecasts project that the energy demand will also continue to grow at high rates in developing countries such as Asia, Africa, and the Middle East. Growth will probably continue at lower rates throughout the industrialized world, particularly in the United States.

This growth will probably push MENA oil and gas exports to the limits of the capacity that MENA countries can reach for several decades. For all the talk of new U.S. energy policies, and energy discoveries in other areas of the world, there is little prospect that the global economy can find any near- to midterm sources of energy that can prevent a growing reliance on MENA energy exports. Proven MENA oil and gas reserves remain too high a percentage of the world's total supply and are much cheaper to produce than most alternative sources of supply, and virtually all major energy forecasts call for a sharp increase in global dependence on MENA exports between 2005 and 2030.

The rise in global demand for energy has illustrated just how important MENA exports will be in the future, and how critical increases in MENA production capacity and exports will be to keeping prices moderate. The MENA region has more than 60 percent of the world's proven conventional oil reserves and is the only region that can be counted upon to increase its production and export capacities in the midterm and the long-term. It has 40 percent of the world's proven gas reserves, and the demand for gas imports is rising sharply. There is no obvious substitute to a rising global dependence on MENA energy for at least the next quarter century.

These trends are forcing energy analysts to rethink the way in which they analyze MENA energy developments. In the past, many analysts and modelers have simply assumed that MENA countries would increase their export capacity to meet world demand at moderate prices. They have not analyzed what MENA countries actually planned to do or the real-world limits to how much and how rapidly they can actually increase production capacity.

Such analysis may have been acceptable as long as there was a significant surplus in existing production capacity relative to demand, although it always creates false expectations on the part of those who confused the projections of models with reality. Today, the issue is what will actually happen: what individual MENA countries plan to do and whether they can do it.

This has led some to talk about an energy crisis in the Middle East. It has led to growing speculation about whether MENA states can really provide the kind of energy exports that current models call for. It has also led others to conspiracy theories about the role MENA states and OPEC (Organization of the Petroleum Exporting Countries) play in world energy markets, fears of U.S. invasions or efforts to control MENA resources, or similar fears of Chinese and Indian efforts to secure energy supplies in ways that bypass global markets.

This book lays out the current facts that shape both MENA capacity to supply energy exports and the possible causes of major interruptions in supply and failures to maintain and expand export capacity at the level the world can really expect. It does not predict a major energy crisis, but it does describe a range of factors that could produce one.

At the same time, it finds that much of the talk about an energy crisis is based more on unrealistic forecasts and expectations than problems in the Middle East. The world needs to adapt to the true nature of the increases in supply that MENA states can be expected to make. The following chapters show that there may be much more of a crisis in unrealistic expectations about energy prices, and unrealistic projections of future export levels, than one in MENA export capacity.

In any case, it is unrealistic to talk about an energy crisis in the Middle East. If an energy crisis exists, it is a global crisis in adjusting to limits on global oil and gas export capacity, higher prices, and the need to find alternatives in the form of different forms of energy, increased efficiency in energy use, and better efforts to conserve energy resources.

REAL-WORLD EXPECTATIONS AND REAL-WORLD VULNERABILITIES

If the world is to develop real-world energy policies, it must set real-world expectations. It must accept the fact that MENA states do face limits on their future energy export capacity and will act according to their perceptions of their own economy interests and not some theoretic model of global need. Energy importers cannot afford to go on living in a climate of illusion about how much energy the MENA region will export in the future, or view limits to actual export capacity in terms of conspiracies by OPEC or MENA states. Living in a paranoid fantasy land is not going to solve anyone's problems.

More is involved than determining what levels of MENA and other global energy production and export capacity are technically feasible and will be provided at future market prices. Virtually all exporting states have some vulnerability to interruptions in their exports and face uncertainties over how much they can maintain and expand export capacity. The geostrategic stability of the MENA region does remain tenuous.

The rise of Islamist extremism and terrorism poses some degree of threat to virtually every nation in the region. Terrorist violence or low-level insurgency is taking place in such key MENA states as Algeria, Egypt, and Saudi Arabia. The Iraq War has not decreased the importance of the Gulf region or the West's reliance on MENA's energy sources. On the contrary, the insurgency in Iraq has highlighted the uncertainty consumers of energy must deal with in the short-, medium-, and long-terms. Iranian and other efforts to acquire weapons of mass destruction threaten to destabilize the region and pose a particular threat to the Gulf. Religious extremism, the Iraq War, the Israeli-Palestinian conflict, and tensions in Lebanon and Syria threaten to divide Islam and create new tensions among Muslim, Christian, and Jew.

Virtually all of the oil-exporting nations in the MENA region face major internal problems with population growth, in creating jobs for a steadily expanding young labor force, in economic reform, in dealing with social changes, and in political modernization. Moreover, the last few years have made it all too clear that the strategic uncertainty in the MENA region can interact with equal levels of uncertainty in Africa, Russia, and Latin America. The MENA region may have the largest proven petroleum and gas reserves, but the risks involved are global.

ORGANIZATION OF THE ANALYSIS AND KEY SOURCES OF DATA AND ESTIMATES

Chapter 1 highlights the importance of the region to the energy market, analyzes trends in global oil and gas supply and demand, and studies the changing nature of global energy dependence. The following chapters analyze each of the key areas impacting the dynamics of energy developments in the MENA region:

- Chapter 2 analyzes key areas of internal risks in the region and general security trends. No energy analysis can be credible without taking into account the security and geopolitical uncertainty faced by countries in the MENA region.

- Chapter 3 analyzes the all too important link between regional stability and economic and political dimensions in the region. It outlines important developments in the MENA countries and the general long-term trends that must be addressed and taken into account in the Middle Eastern and North African energy sector.

- Chapter 4 analyzes energy development in the Gulf. It provides country-by-country analysis, national oil and gas production capacity trends, future plans, and energy risk assessment by country. It covers Bahrain, Iran, Iraq, Qatar, Oman, Saudi Arabia, and Yemen.

- Chapter 5 analyzes development in the Levant countries of Jordan, Lebanon, Palestine, Israel, Egypt, and Syria. It covers the national plans and provides country-by-country assessment of energy risks and development in their oil and gas industries

- Chapter 6 analyzes national oil and gas production capacity trends, future production expansion plans, and the overall energy risk for the North African nations of Algeria and Libya. It also provides a general overview of developments in North Africa.

- Chapter 7 focuses on methods of energy analyses, the importance of transparency, and of developing credible forecasting models. It compares different methods and energy reports by agencies such as the IEA (International Energy Agency), EIA (Energy Information Administration), and OPEC. The chapter also studies how the lack of robust modeling is contributing to the aura of uncertainty in the energy market, and it examines important issues in Middle Eastern energy.

- Chapter 8 studies the future of MENA energy supply and demand, the importance of foreign and domestic investment, and the potential impact technological and best practices can have on the global and Middle Eastern energy balance.

- Chapter 9 focuses its analyses on broad scenarios affecting the future of the region and the energy market. It puts current energy developments, announced country plans, and key geopolitical forces in perspective and examines their impact on policy and strategic planning.

This analysis draws upon both national data and two key official efforts at energy analysis and modeling. There are many sources of global energy estimates, they use many different models, and their results differ in detail. There are also many major uncertainties as to the size of the oil and gas reserves in any given country, the cost of extracting them, future energy demand, future energy costs, and virtually every other aspect of energy analysis and forecasting.

Most energy experts do, however, make use of the data and modeling produced by the EIA of the U.S. Department of Energy (DOE) and the IEA. The forecasts and estimates of the EIA and the IEA also represent one of the few modeling efforts that receive public review and which are supported by large analytic resources and a reasonable level of transparency. The EIA also annually recalibrates its forecasts and estimates based on its past degree of accuracy.

As Chapter 7 shows, there are serious limits to these forecasts. In effect, they estimate the amount of energy the global economy would like and then find ways to show how energy supplies could increase to meet market needs without becoming unaffordable. Estimates by the EIA and IEA are, however, key benchmarks most analysts use to analyze the global energy market. They also are probably all too realistic in showing just how critical a role future demand for oil and gas will play in the world's overall use of energy, in spite of the impact of other sources of energy, advances in technology, advances in energy efficiency, and the impact of conservation.

There is never any certainty to estimates of the future role of given sources of energy. It is always possible that some massive technological breakthrough will occur that will sharply reduce the need for oil or some massive new source of energy resources will be found outside the Middle East. However, ever since the United States first sought to reduce its dependence on foreign oil—which took place as part of Project Independence, beginning in the early 1970s—various experts have promised the solution could come from offshore oil reserves outside the MENA region or from other sources of energy like fuel cells, tar sands, shale oil, nuclear power, fusion, geothermal energy, biomass, wind, conservation, and a host of other means.

None of these promises has succeeded in altering the fundamental balance of world energy supplies or in reducing global economic dependence on exports from the MENA region. In fact, despite 30 years of efforts to find substitutes for MENA energy resources and exports, estimates of the percentage of the world's proven oil reserves located in the MENA area have increased, as have the projections of global dependence on MENA exports.

OIL AND GAS VS. OTHER SOURCES OF ENERGY

No analysis of the present and the future impact of MENA oil and gas exports can be decoupled from an assessment of the overall trends in global energy balances. Any such analysis can only be illustrative, given the uncertainties involved, and recent EIA and IEA projections present the problem that they have not comprehensively examined the impact of sustained high oil prices. Nevertheless, such projections are probably broadly correct in showing how demand for petroleum interacts with other forces of energy supply.

Such trends also become clear only when they are both quantified and portrayed in graphic form. They are too complex to describe in prose alone, and the interactions between the trends involved can be understood only by taking the time to compare the numbers and trends developed in different estimates and projections. Like many areas of economics, this may make energy analysis a "dismal science" for those who dislike numbers. However, there is no other way to approach energy analysis and to properly illustrate both what estimates predict and the major uncertainties in such estimates.

Global Energy Trends and the Future Role of Oil and Gas

The following charts and tables provide an overview of the possible trends in world energy supplies and consumption.

• Figure 1.1 shows an estimate by the EIA of world energy consumption by type of fuel between 1970 and 2025. It projects major increases in the use of hydroelectric and renewable energy (shown as "other") by 116 percent from 1970 to 1990, 24 percent from 1990 to 2000, and 49 percent from 2003 to 2025. It also projects 108 percent from 1970 to 1990 and 64 percent from 2003 to 2025 in the use of natural gas. In addition, the world's use of coal will increase by 52 percent from 1970 to 1990, 11 percent from 1990

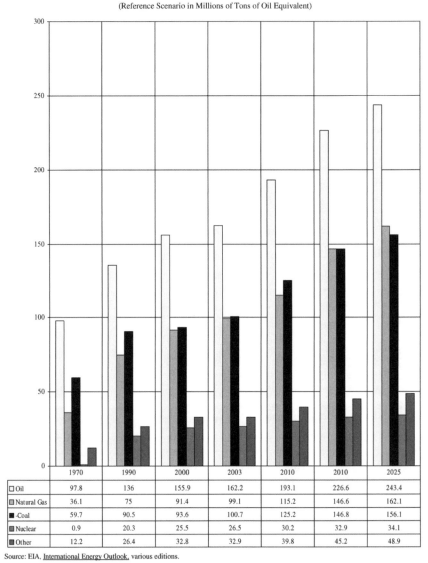

(Reference Scenario in Millions of Tons of Oil Equivalent)

	1970	1990	2000	2003	2010	2010	2025
☐ Oil	97.8	136	155.9	162.2	193.1	226.6	243.4
▨ Natural Gas	36.1	75	91.4	99.1	115.2	146.6	162.1
■ -Coal	59.7	90.5	93.6	100.7	125.2	146.8	156.1
▨ Nuclear	0.9	20.3	25.5	26.5	30.2	32.9	34.1
▨ Other	12.2	26.4	32.8	32.9	39.8	45.2	48.9

Source: EIA, International Energy Outlook, various editions.

Figure 1.1   EIA Projection of World Primary Energy Consumption by Type of Fuel

to 2003, and 55 percent from 2003 to 2025. The world still, however, nearly doubles its use of oil and will see an increase in oil by 50 percent from 2003 to 2025.

• Figure 1.2 shows a similar projection to 2030 by the IEA. From 2003 to 2030, hydro increases by nearly 62 percent, coal by 44 percent, gas by 76 percent, nuclear by 12 percent, biomass by 15 percent, and renewables by 404 percent. Despite the growth in other sources of energy, the demand for oil will grow by 47 percent.

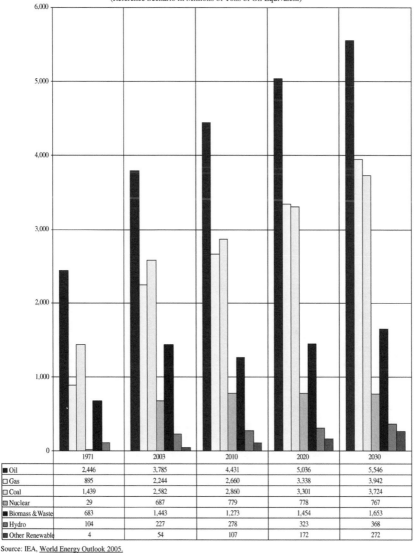

(Reference Scenario in Millions of Tons of Oil Equivalent)

	1971	2003	2010	2020	2030
■ Oil	2,446	3,785	4,431	5,036	5,546
□ Gas	895	2,244	2,660	3,338	3,942
□ Coal	1,439	2,582	2,860	3,301	3,724
▨ Nuclear	29	687	779	778	767
■ Biomass & Waste	683	1,443	1,273	1,454	1,653
■ Hydro	104	227	278	323	368
■ Other Renewable	4	54	107	172	272

Source: IEA, World Energy Outlook 2005.

Figure 1.2   IEA Projection of World Primary Energy Consumption by Type of Fuel

• Figure 1.3 shows that the EIA and the IEA both model a significant drop in the future rate of increase in oil consumption relative to the past, but that both estimate there will still be a 1.7–1.8 percent annual increase in oil use over the next 22–27 years.

The actual future will, of course, be different. A quick glance at the details of Figures 1.1 and 1.2, or the comparison in Figure 1.3, shows that the EIA and the IEA project strikingly different futures for renewables and nuclear energy, and significant differences in the relative growth of gas and coal. These estimates also understate the

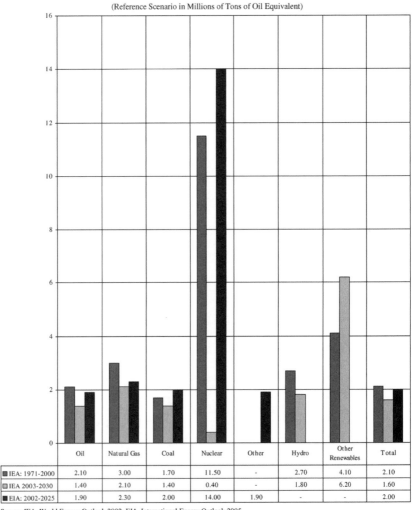

(Reference Scenario in Millions of Tons of Oil Equivalent)

	Oil	Natural Gas	Coal	Nuclear	Other	Hydro	Other Renewables	Total
■ IEA: 1971-2000	2.10	3.00	1.70	11.50	-	2.70	4.10	2.10
▨ IEA 2003-2030	1.40	2.10	1.40	0.40	-	1.80	6.20	1.60
■ EIA: 2002-2025	1.90	2.30	2.00	14.00	1.90	-	-	2.00

Source: IEA, World Energy Outlook 2002; EIA, International Energy Outlook 2005.

Figure 1.3 Comparative EIA and IEA Estimates of Average Annual Percentage of Future Growth of World Energy Consumption by Type of Fuel: 1971–2030

uncertainties involved. As has been touched upon earlier, these estimates are based on demand-driven modeling techniques that tend to assume incremental supply is available at moderate prices. Even if they were more realistic in estimating how quickly supply will increase, real-world historical trends are never smooth or consistent. No one can predict a technological breakthrough. No one can predict economic growth or environmental developments with any precision, and extrapolating existing trends in known sources of energy over more than two decades is certain to produce substantial errors.

Oil and Gas Dependence vs. Dependence on Other Forms of Energy

At the same time, these projections do help illustrate the fact that energy risk goes far beyond oil and gas and the MENA region. The highest risks in terms of the gains in the various projections of future sources of energy do not seem to be in oil and gas supply, but rather in nuclear energy (because of the perceived safety risk), and in the use of coal (environmental problems). While renewables are often seen as desirable in terms of emissions, virtually all of the projected gain in these projections comes from large hydroelectric plants, which are increasingly seen as posing major environmental risks of a different kind.

Put differently, if any shortfall occurs in the other areas of global energy supply, the demand for petroleum will actually be much higher than models presently estimate, particularly because oil remains by far the most efficient way of transporting energy flexibly over long distances. Similarly, the higher the rate of global economic growth and the more developing nations actually develop, the higher the demand for oil and oil imports.

The world's economy does, however, have tremendous momentum. Drastic shifts in the global balance of different types of energy supply involve massive investments in production, transportation, and end-use equipment that are expensive and difficult to accomplish. The world is hard to change in broad structural terms, and most shifts in energy are costly; availability, export methods, and technology are incremental and take decades to have a major global impact.

In practice, this means that fossil fuel will be the single most important energy source of meeting the global energy needs for the next three decades. Recent studies by the EIA and the IEA assert that fossil fuel will continue to dominate the world energy market at least to 2030. The IEA estimates the following:[1]

Fossil fuels will continue to dominate energy supplies, meeting more than 80% of the projected increase in primary energy demand. Oil remains the single most important fuel, with two-thirds of the increase in oil use coming from the transport sector. Demand reaches 92 mb/d in 2010 and 115 mb/d in 2030. The lack of cost-effective substitutes for oil-based automotive fuels will make oil demand more rigid. Natural gas demand grows faster, driven mainly by power generation. It overtakes coal as the world's second-largest primary energy source around 2015. The share of coal in world primary demand falls a little, with demand growth concentrated in China and India. The share of nuclear power declines marginally, while that of hydropower remains broadly constant. The share of biomass declines slightly, as it is replaced with modern commercial fuels in developing countries. Other renewables, including geothermal, solar and wind energy, grow faster than any other energy source, but still account for only 2% of primary energy demand in 2030.

The EIA reference case estimated total world oil consumption increased from 66.5 million barrels per day (MMBD) in 1990, and 78.0 MMBD in 2001, to 78.2 MMBD in 2002. This case projected total consumption to reach 94.6 MMBD in 2010, 103.2 MMBD in 2015, 111.0 MMBD in 2020, and 119.2 MMBD in 2025. While this is only an average annual increase of 1.9 percent per year from 2002 to 2025, it would amount to a total increase of 41.0 MMBD during the forecast period—a cumulative increase of 52 percent.[2]

It seems doubtful that any of the forces now at work could produce major short-term (2006–2010) changes in the broad structure of global energy balances, and there are many reasons why the Middle East will probably continue to dominate the world oil market for the next two decades even if substantial changes were to take place in global demand.

The MENA area has been, and will continue to be, a critical factor in meeting global demand and in providing oil exports for simple and straightforward reasons. It has more oil, its oil is cheaper to produce, and it has the infrastructure to export energy cheaply and in large amounts. Its cost for additional production is low by comparative standards, and domestic demand for oil in the MENA region is low relative to total production capacity. In fact, if some major breakthrough in other sources of energy or conservation should reduce global demand for oil, higher cost producers in other areas would probably have to cease producing or reduce production first, meaning that the MENA region would retain its importance in the world market.

THE IMPORTANCE OF MENA OIL TO THE GLOBAL MARKET

The MENA region dominates world oil exports today and will almost certainly do so for decades to come. This is true even if one assumes steady progress in conservation, major improvements in the supply of renewables, and major increases in energy supplies from gas, coal, nuclear power, and renewables.

Recent estimates indicate that the region has well over 60 percent of the total world oil proven reserves (733.9 billion barrels), and it produced 24.57 MMBD or 24 percent of the world's total oil production. Even if Canadian tar sands are included, which are estimated to contain 175 billion barrels, the MENA region still has 53 percent of the world total oil reserves. In addition, the Middle East contains over 40 percent of the world gas reserves or 2,570 trillion cubic feet (Tcf).[3]

Current Estimates of Proven MENA Oil Reserves vs. the World Total

One can argue the validity of the way that proven oil reserves are estimated. Some producers have inflated their "proven" reserves to project strategic importance, which has added to the uncertainty and the lack of transparency. Limited hard data are available to validate many current national claims.

Nevertheless, most sources estimate that most of the world's proven oil and gas reserves are in the Middle East and North Africa. For example, the IEA summarized the importance of MENA energy in its *World Energy Outlook 2005* as follows:[4]

> The greater part of the world's remaining reserves lie in that region [the MENA]. They are relatively under-exploited and are sufficient to meet rising global demand for the next quarter century and beyond. The export revenues they would generate would help sustain the region's economic development. But there is considerable uncertainty about the pace at which investment in the region's upstream industry will occur, how quickly production capacity will expand and, given rising domestic energy needs, how much of the expected increase in supply will be available for export. The implications for both MENA producers and consuming countries are profound.

British Petroleum (BP) is widely quoted as a credible source of global oil and gas reserves, and its estimates of world's oil resources, and MENA's share, reflect the following trends:[5]

- In 1980, the world had a total of 667.1 billion barrels of proven oil reserves, and the Middle East and North Africa accounted for 396.0 billion barrels of this total (59.3 percent).
- In 1988, the world had a total of 996.2 billion barrels of proven oil reserves, and the MENA area accounted for 689.5 billion barrels of this total (69.2 percent).
- In 1998, the world had a total of 1,069.5 billion barrels of proven oil reserves, and the MENA area accounted for 719 billion barrels of this total (67.2 percent).
- In 2002, the world had a total of 1,047.7 billion barrels of proven oil reserves, and the MENA area accounted for 783.8 billion barrels of this total (66.5 percent).
- In 2004, the world had a total of 1,188.6 billion barrels of proven oil reserves, and the MENA area accounted for 795.3 billion barrels of this total (67 percent).

As is the case with many areas of energy analysis, estimates do not always agree on fundamental issues of production, reserves, and export capacity. What virtually all sources do agree upon, however, is the importance of oil and natural gas and oil in the MENA region to the global energy market.

It is also important to understand that the vast majority of these reserves are held in the Gulf. If one uses the BP estimates, the Gulf and Yemen have 61.4 percent of the world's reserves, the Levant has 0.6 percent, and North Africa has 4.5 percent.[6] Moreover, if one uses the conventional method of estimating proven oil reserves, the broad patterns in the distribution of the world's oil reserves by country have not changed in more than a decade. In fact, unless one counts recent efforts to reclassify Canadian tar sands as part of proven oil reserves, the end result of more than 30 years of exploration since the oil embargo of 1973 has been to increase the Middle East's percentage of proven total world oil reserves.

The Future Impact of MENA Oil Reserves

There are a number of different ways to estimate oil reserves, and there are many debates over the size of probable oil reserves and future discoveries, over how to count heavy oil and tar sands, and over the impact the rate of future advances in recovery technology will have on estimates of exploitable reserves. The IEA currently defines the types of reserves as follows:[7]

- Proven reserves: the hydrocarbons that have been discovered and for which there is reasonable certainty that they can be extracted profitably.

- Probable reserves: the volumes that are thought to exist in accumulations that have been discovered and that are expected to be commercial, but with less certainty than proven reserves.

- Possible reserves: the volumes in discovered fields that are less likely to be recoverable than probable reserves.

The U.S. Geological Survey has two more types of reserves and defines them as follows:

- Known reserves: discovered crude oil accumulations that are considered economically viable to produce.

- Undiscovered reserves: quantities of crude oil that geological data and engineering information indicate exist outside known oil fields.

Uncertainty and Risk in Reserve Estimates

As has been discussed earlier, estimates of the share of the world's total oil reserves that are in the MENA area have increased steadily for a quarter of a century, but some experts question how realistic current estimates of proven and potential reserves really are, and how long the gains from new exploration, drilling, and production technologies can be sustained. There are serious debates over how given countries are characterizing and managing their oil reserves, and these involve such key MENA countries as Iraq, Kuwait, Saudi Arabia, and the United Arab Emirates (UAE). Matthew R. Simmons, for example, has challenged current calculations of Saudi oil reserves, especially in the Kingdom's ability to manage reservoirs and replace aging giant and supergiant oil fields.[8]

Chapter 4 examines the Simmons challenge and the Saudi Aramco response in detail, but it is worth noting that oil is a strategic commodity, and many oil-producing nations have used their estimates of reserves to stress their strategic importance. There was a race among Gulf States to increase their claims to proven reserves during the Iran-Iraq War, both to obtain outside aid and to gain political status. Kuwait, for example, claimed its reserves suddenly jumped from estimates of around 65.4 billion barrels in the early 1980s (1982), to around 90.0 billion in 1985. Iran claimed an increase from 58.0 billion barrels in 1982 to 100.0 billion in 1987.

Abu Dhabi increased its claims from 58.0 billion to 92.9 billion during this same period. Iraq responded with claims that its reserves increased from 31.0 billion in 1982 to 100.0 billion in 1988, and Saudi Arabia increased its claims from 163.4 billion in 1982 to 257.5 billion in 1989.

Even when politics are not an issue, there are enough inevitable uncertainties in the data and methodology, resulting in many different estimates of proven, probable, known, discovered, and undiscovered oil reserves. The most notable open source estimates are the *Oil and Gas Journal,* the BP annual *Statistical Review of World Energy,* the *U.S. Geological Survey,* and the *OPEC Annual Review.*

The most quoted type of reserves is proven, and the IEA has described the different attempts to estimate proven oil reserves as follows:[9]

> The most widely-quoted primary sources of global reserves data—the Oil and Gas Journal (O&GJ), World Oil, Cedigaz (for gas), and the US Geological Survey (USGS)—compile data from national and company sources. In addition, OPEC compiles and publishes data for its member countries. IHS Energy (formerly Petroconsultants) also compiles data but for "proven plus probable" reserves only. Other organizations, including BP, publish their own global estimates based mainly on data from the main primary sources. Despite the differences in approaches among these organizations, current estimates of remaining proven reserves worldwide do not vary greatly. BP puts global oil reserves at the end of 2003 at 1 148 billion barrels. Other estimates for the same year range from World Oil's 1 051 billion to IHS Energy and O&GJ's 1 266 billion barrels. Both BP and O&GJ include Canadian oil sands which increases their estimates substantially, although differences in their approaches produce very different results: BP includes 11 billion barrels of oil-sands reserves under active development. O&GJ includes all proven oil-sands reserves, which they estimate at 174 billion barrels. World Oil excludes natural gas liquids and Canadian oil sands. Differences among all the main sources are small in terms of the number of years of remaining reserves: 36 according to World Oil and 44 according to O&GJ.

Most estimates are close to each other. The differences tend to stem more from the definition of what is proven, probable, known, and undiscovered than from major differences in the input data. As the following figures show, estimates tend to agree more often than not:

- Figure 1.4 compares the BP estimates of proven oil reserves by region to that of the USGS 2000 of unknown and discovered oil reserves. It shows that the Middle East reserves (excluding North Africa) have 62 percent of the world's proven oil reserves, 42 percent of the known oil reserves, and 31 percent of the undiscovered oil reserves. In each case, it is greater than any other region in the world.

- Figure 1.5 shows a BP estimate of the trends in Middle Eastern reserves over the last two decades and shows an upward trend in each country. It also shows how Middle Eastern nations rank relative to other leading oil producers in terms of total proven reserves. From 1984 to 2004, the BP estimates of proven oil reserve in Iran increased by 125 percent, Iraq by 77 percent, Kuwait by 7 percent, Qatar by 338 percent, Saudi Arabia by 53 percent, Syria by 129 percent, the United Arab Emirates by 201 percent, Yemen by 2,800 percent, Algeria by 31 percent, and Libya by 83 percent. Egypt and

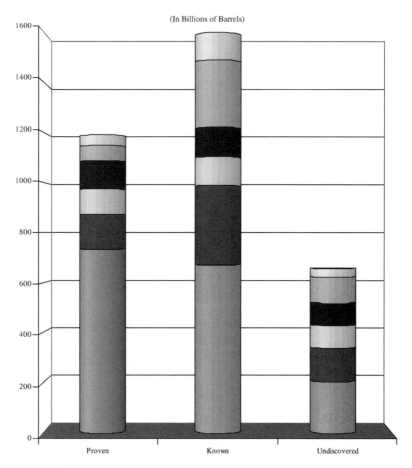

	Proven	Known	Undiscovered
☐ Asia Pacific	41.1	107.7	32.1
☐ North America	61.0	268.0	106.3
■ Africa	112.2	116.5	89.7
☐ S & C America	101.2	113.5	91.1
■ Europe & Eurasia	139.2	315.7	132.8
☐ Middle East	733.9	672.6	204.8

Sources: BP, Statistical Review of World Energy 2005, and the USGS 2000.

Figure 1.4   BP and USGS 2000 Estimates of World Oil Resources in 2004

Tunisia's proven reserves estimates decreased by 10 percent and 67 percent, respectively. Total MENA proven oil reserves increased by approximately 69 percent during the last two decades.

• Figure 1.6 compares the trend in the Middle Eastern proven oil reserves to that of the rest of world. The largest growth was seen in South and Central America, which increased its proven oil reserves by 179 percent. Africa follows with a 94-percent increase, the Middle

Figure 1.5   BP Estimates of MENA and Other Regional Proven Oil Reserves 1984–2004 (in Billions of Barrels)

Nation	1984	1994	2004	% Change 1984–2004	% of World Total
Bahrain	N/A	N/A	N/A	N/A	N/A
Iran	58.90	94.30	132.50	125%	11.10
Iraq	65.00	100.00	115.00	77%	9.70
Kuwait	92.70	96.50	99.00	7%	8.30
Oman	3.90	5.10	5.60	44%	0.50
Qatar	4.50	3.50	15.20	238%	1.30
Saudi Arabia	171.70	261.40	262.70	53%	22.10
Syria	1.40	2.70	3.20	129%	0.30
UAE	32.50	98.10	97.80	201%	8.20
Yemen	0.10	0.10	2.90	2,800%	0.20
Other	0.20	0.10	0.10	–50%	less than 0.050
Total Middle East	**430.80**	**661.70**	**733.90**	**70%**	**61.70**
Algeria	9.00	10.00	11.80	31%	1.00
Egypt	4.00	3.90	3.60	–10%	0.30
Libya	21.40	22.80	39.10	83%	3.30
Tunisia	1.80	0.30	0.60	–67%	0.10
Total MENA	**467.00**	**698.70**	**789.00**	**69%**	**66.40**
Other Africa	21.60	28.00	57.10	164%	4.70
Asia Pacific	38.10	39.20	41.10	8%	3.50
Europe/Eurasia	96.70	80.30	139.20	44%	11.70
North America	101.90	89.80	61.00	–40%	5.10
S. & C. America	36.30	81.50	101.20	179%	8.50
World Total	**761.60**	**1017.50**	**1188.60**	**56%**	**100.00**

Source: BP, *Statistical Review of World Energy 2005*. N/A = not available.

East with a 70-percent increase, and Europe and Eurasia increased by 40 percent. The only decrease occurred in North America, which saw a 40-percent decline in its proven oil reserves.

The Canadian Tar Sands Debate

Most significant departures from the conclusions reached in the BP estimate reflect the decision to include Canadian tar sands in estimates of world reserves. Canada estimates that such tar sands total 175 billion barrels of oil and has proposed they be included in the estimate of proven oil reserves on the grounds that they can be produced at a cost of $16 to $26 per barrel, less transportation.

The USGS indicated in 2000 that recoverable tar sands could be only 20 to 33 percent of what the Canadian Energy Board claims.[10] The EIA estimates that if such tar sands are included in the pool of proven reserves, the MENA share of world reserves would be reduced from some 65 percent to around 57 percent. The EIA analysis also

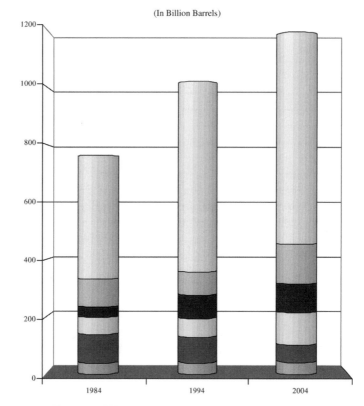

(In Billion Barrels)

	1984	1994	2004
☐ Middle East	430.80	661.70	733.90
☐ Europe & Eurasia	96.70	80.30	139.20
■ S&C America	36.30	81.50	101.20
☐ Africa	57.80	65.00	112.20
■ N. America	101.90	89.80	61.00
☐ Asia Pacific	38.10	39.20	41.10

Note: the Middle East excludes North Africa. If one includes North Africa, the proven oil reserves will be 467.0, 698.7, and 789.0 in 1984, 1994, and 2004 respectively.

Source: BP, Statistical Review of World Energy 2005.

Figure 1.6   World Proven Oil Reserves Trends by Region: 1984–2004

indicates that the commercialization of Canadian tar sands at this price spread may prove to be commercial over time, but that this will take years to fully conform and requires a massive new production and transportation infrastructure.

Near- to midterm production capacity is also an issue in evaluating the importance of tar sands. The U.S. DOE estimates that even if affordable production of this volume of reserves of tar sands does prove valid in real-world economic terms, it will lead to only 2.2 MMBD worth of actual production by 2025, and 1.0 MMBD of exports to the United States.[11] This makes tar sands an interesting possibility, but one with limited short- to midterm impact.

USGS Estimates of Potential and Probable Reserves

The U.S. Geological Survey provides another way of considering how estimates of world reserves might change in the future. It not only estimates proven reserves—which are recoverable with today's technology and today's costs—but the potential growth in reserves in known fields and the probable size of undiscovered fields. According to the USGS, the present total size of proven reserves is 1,212.9 billion barrels compared to 1,188.6 billion barrels of oil. The Middle East has 685.64 billion barrels, or 58 percent of the total.[12]

If one looks at potential discoveries through 2025, the USGS estimates that known reserves and fields will be found to have another 730.5 billion barrels by 2025 and that the Middle East will then have 252.5 billion barrels, or 34.6 percent of these new discoveries. If one combines proven reserves and reserve growth, the Middle East would have a total of 938.1 billion barrels, or 48 percent of 1,943 billion barrels. This indicates that the Middle East could shrink as a percent of future world production after 2025.

This would be even truer if one considers the USGS estimate of undiscovered fields and reserves. The USGS estimates that undiscovered fields and reserves could amount to another 939.9 billion barrels and that the Middle East could have 269.2 billion barrels, or 28.7 percent of this total. If one combines proven reserves, reserve growth, and undiscovered reserves, the Middle East would have 1,207.3 billion barrels, or 42 percent of a global total of 2,882.9 billion barrels.

In short, the Middle East would remain of critical strategic importance, but could lose its present level of dominance at some point between 2020 and 2030. This is an important long-term possibility, but one with little practical importance for current and midterm energy policy.

The IEA Estimate of Reserves

Figure 1.7 compares the 2002 International Energy Agency estimates of remaining, undiscovered, and total oil production of major oil production nations. It compares how much each country has produced to date to that of its remaining reserves in billion barrels of oil. The IEA uses a mixture of its own databases and the USGS estimates and calculates total world oil production up to 2002 at 718.0 billion barrels and annual production in 2001 at 75.8 MMBD. It projects 959.0 billion barrels of remaining reserves and 939 billion barrels of undiscovered reserves.

Saudi Arabia is estimated to have an estimated 221 billion barrels in remaining reserves and 136 billion barrels in undiscovered reserves. Russia ranks second with 137 billion barrels in remaining reserves and 115 billion barrels in undiscovered reserves. Other Middle East states dominate the rest of the picture.[13] According to the EIA, the world oil production averaged 80.9 MMBD from 2001 to the end of

Figure 1.7   IEA Estimates of Remaining Oil Reserves, Undiscovered Resources, and Total Production in 2002 (in Billions of Barrels)*

Rank	Country	Remaining Reserves	Undiscovered Resources	Total Production
1	Saudi Arabia	221.0	136.0	73.0
2	Russia	137.0	115.0	97.0
3	Iraq	78.0	51.0	22.0
4	Iran	76.0	67.0	34.0
5	United Arab Emirates	59.0	10.0	16.0
6	Kuwait	55.0	4.0	26.0
7	United States	32.0	83.0	171.0
8	Venezuela	30.0	24.0	46.0
9	Libya	25.0	9.0	14.0
10	China	25.0	17.0	24.0
11	Mexico	22.0	23.0	22.0
12	Nigeria	20.0	25.0	4.0
13	Kazakhstan	20.0	25.0	4.0
14	Norway	16.0	23.0	9.0
15	Algeria	15.0	10.0	10.0
16	Qatar	15.0	5.0	5.0
17	United Kingdom	13.0	7.0	14.0
18	Indonesia	10.0	10.0	15.0
19	Brazil	9.0	55.0	2.0
20	Neutral Zone	8.0	0.0	5.0
21	Others	73.0	220.0	91.0
Total		**959.0**	**939.0**	**728.0**

*Note: Estimates include crude oil and NGLs; estimates are taken from the IEA and USGS databases.

2005, which means that the world oil reserves are declining on average by that level.

If one looks at the IEA estimate of the reserves of other major MENA oil producers, they have the following resources:

- Iraq is estimated to have 78 billion barrels in remaining reserves and 61 billion barrels in undiscovered reserves. (Estimates of Iraq's oil reserves and resources vary widely since only 10 percent of the country's resources have been explored. Various reports—The Baker Institute, Center for Global Energy Studies, the Federation of American Scientists—indicated that the main deep oil-bearing formations of Iraq located in the Western Desert region could contribute additional resources up to 100 million barrels. These resources, however, have not been explored.)

- Iran is estimated to have 78 billion barrels in remaining reserves and 67 billion barrels in undiscovered reserves.

- The United Arab Emirates is estimated to have 59 billion barrels in remaining reserves and 10 billion barrels in undiscovered reserves.

- Kuwait is estimated to have 55 billion barrels in remaining reserves and 4 billion barrels in undiscovered reserves.
- Libya is estimated to have 25 billion barrels in remaining reserves and 9 billion barrels in undiscovered reserves.
- Algeria is estimated to have 15 billion barrels in remaining reserves and 10 billion barrels in undiscovered reserves.
- Qatar is estimated to have 15 billion barrels in remaining reserves and 5 billion barrels in undiscovered reserves.
- The Kuwaiti-Saudi Neutral Zone is estimated to have 8 billion barrels in remaining reserves and 0 billion barrels in undiscovered reserves.
- The United States is estimated to have 32 billion barrels in remaining reserves and 83 billion barrels in undiscovered reserves.

Cost Factors and Estimates of Reserves

Other factors need to be considered in evaluating such estimates of near- and mid-term impact of new discoveries on the world oil market. The cost of production from outside the MENA region varies sharply from region to region once one considers reserve growth and undiscovered reserves. Much of the production would have to come from the former Soviet Union, and from Latin American and African states, where production costs are often at least twice those in the Middle East. The estimates of reserve growth require major advances in enhanced oil recovery to make production economically viable outside the Middle East, and it can take decades to create the production and export infrastructure necessary to exploit undiscovered reserves.

MENA Oil Production and Production Capacity

Given these factors, it is hardly surprising that most estimates indicate that the MENA region will steadily expand its oil production, increase its share of world production, and increase its impact on the global economy through 2025–2030. There are major uncertainties in such estimates, and it must again be stressed that they are based upon demand-driven models that can exaggerate the ease with which major long-term increases can be made in supply at moderate prices. Despite these uncertainties, projections by the EIA and the IEA remain the best benchmarks for analysts and policy makers.

Production Levels vs. Sustainable Production Capacity

Given the tightness of the global oil market, production capacity is what matters most in determining market prices. It represents the oil-producing nations' abilities to meet sudden demand surges or supply disruptions, or, put differently, it represents the market supply ability to meet demand at a given price.

Production capacity is harder to measure than actual production, and sustainable production capacity is even harder. The ongoing debates about surplus capacity and future capacity make it all too clear that the global energy market depends on the ability of producers to sustain a given level of capacity for a given period of time. It is important, therefore, to start with some general definitions of these metrics:

- Production level: represents the actual amount of oil being pumped by a given country or region. It is the easiest metric to measure because it measures how much countries export plus how much they consume locally and store in their strategic reserves. While it measures actual levels of oil, it does require a level of transparency from the producing nation that is lacking in most instances. Agencies such as the IEA, the EIA, and OPEC provide credible estimates.

- Production capacity: measures the actual production level plus the ability of the producing nation to instantly increase its oil production level. The problem with measuring capacity levels is that it depends on more than actual levels. Agencies have to rely on the world of the producing nations. In addition, when forecasting future production capacity, modelers have to take into account present and future geopolitical risks, the global economic environment, the impact of technological breakthroughs, and the actual level of resources.

- Sustainable production capacity: represents the level of production capacity that a producing nation can sustain for a certain time. This is hard to estimate and is equally uncertain as a metric because it depends on the definition of "certain time" and "sustainable."

The other important measure, which is derived from the three preceding metrics, is the level or sustainable surplus capacity, which measures how much more a producer can increase and sustain a certain production level. Currently the level of surplus capacity is getting a lot of attention, especially given high oil prices and the ongoing fear of supply disruptions.

Historical Regional Production Patterns: Upward Trends

Historically, OPEC has dominated the production capacity and level of oil-exporting states, and the Gulf-OPEC countries, Saudi Arabia, Kuwait, Iran, Iraq, and Qatar, have dominated OPEC's total production level, capacity, and sustainable capacity. Between 1970 and 2004, the MENA share of total world production ranged from nearly 25 percent to as high as 42 percent.

Figure 1.8 shows the trend of world oil production levels from 1984 to 2004 by region. It shows that the MENA region production doubled between 1984 and 2004. It produced 14.3 MMBD or 24.9 percent of the world's total in 1984, 23.8 MMBD or 35.6 percent in 1994, and 28.8 MMBD or 36 percent in 2004. Other regions' oil production levels increased in a similar manner. Africa's oil production also doubled during the same period. In 1984, it produced 4.87 MMBD and 9.26 MMBD in 2004. North America's production level decreased during the same period from 15.2 MMBD in 1984 to 14.1 MMBD in 2004.

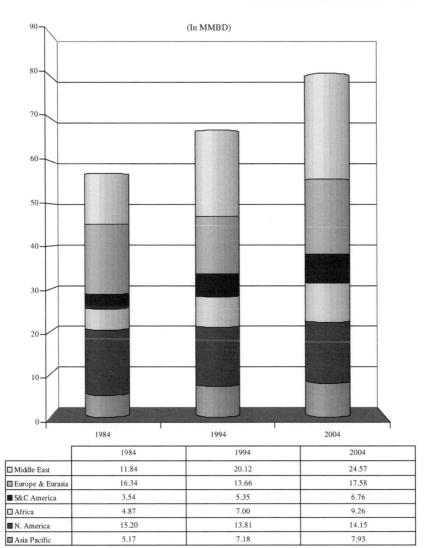

(In MMBD)

	1984	1994	2004
□ Middle East	11.84	20.12	24.57
▣ Europe & Eurasia	16.34	13.66	17.58
■ S&C America	3.54	5.35	6.76
□ Africa	4.87	7.00	9.26
■ N. America	15.20	13.81	14.15
▣ Asia Pacific	5.17	7.18	7.93

Source: BP, Statistical Review of World Energy 2005.

Figure 1.8   World Oil Production Trends 1984–2004

The country-by-country oil production levels in the Middle East show somewhat similar patterns, although important differences exist from country to country. From 1975 to 2004, the production levels of all the MENA countries' except Tunisia, Iraq, and Iran increased, as is shown in Figure 1.9. This pattern of increases somewhat understates the trend in production capacity, however, because of the impact of OPEC quotas, the ongoing requirements imposed on each member, and the impact of fluctuations in oil prices. Saudi Arabia, for example, experienced high levels of production fluctuations between 1970 and 1990. It produced 3.58 MMBD in

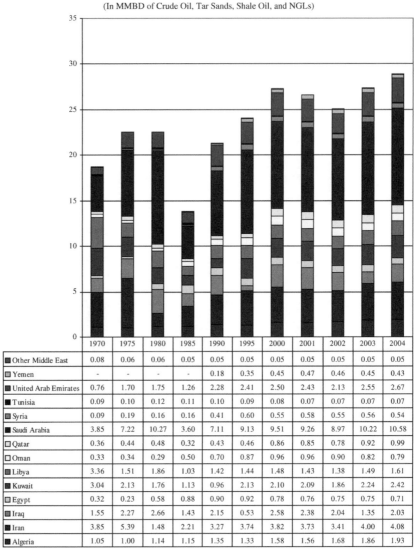

(In MMBD of Crude Oil, Tar Sands, Shale Oil, and NGLs)

	1970	1975	1980	1985	1990	1995	2000	2001	2002	2003	2004
■ Other Middle East	0.08	0.06	0.06	0.05	0.05	0.05	0.05	0.05	0.05	0.05	0.05
▨ Yemen	-	-	-	-	0.18	0.35	0.45	0.47	0.46	0.45	0.43
■ United Arab Emirates	0.76	1.70	1.75	1.26	2.28	2.41	2.50	2.43	2.13	2.55	2.67
■ Tunisia	0.09	0.10	0.12	0.11	0.10	0.09	0.08	0.07	0.07	0.07	0.07
▨ Syria	0.09	0.19	0.16	0.16	0.41	0.60	0.55	0.58	0.55	0.56	0.54
■ Saudi Arabia	3.85	7.22	10.27	3.60	7.11	9.13	9.51	9.26	8.97	10.22	10.58
□ Qatar	0.36	0.44	0.48	0.32	0.43	0.46	0.86	0.85	0.78	0.92	0.99
□ Oman	0.33	0.34	0.29	0.50	0.70	0.87	0.96	0.96	0.90	0.82	0.79
▨ Libya	3.36	1.51	1.86	1.03	1.42	1.44	1.48	1.43	1.38	1.49	1.61
■ Kuwait	3.04	2.13	1.76	1.13	0.96	2.13	2.10	2.09	1.86	2.24	2.42
□ Egypt	0.32	0.23	0.58	0.88	0.90	0.92	0.78	0.76	0.75	0.75	0.71
▨ Iraq	1.55	2.27	2.66	1.43	2.15	0.53	2.58	2.38	2.04	1.35	2.03
■ Iran	3.85	5.39	1.48	2.21	3.27	3.74	3.82	3.73	3.41	4.00	4.08
■ Algeria	1.05	1.00	1.14	1.15	1.35	1.33	1.58	1.56	1.68	1.86	1.93

Source: BP, Statistical Review of World Energy 2005.

Figure 1.9   Middle Eastern Petroleum Production by Country: BP Estimate for 1970–2004

1970, its production level jumped to 10.27 MMBD in 1980, then dropped back to 3.60 MMBD in 1985, and then jumped to 7.11 MMBD in 1990.

The same is true for many of the other MENA countries from 1970 to 2004:

- Algeria produced 1.05 MMBD in 1970, 1.14 MMBD in 1980, 1.35 MMBD in 1990, 1.58 MMBD in 2000, and 1.93 MMBD in 2004.

- Libya produced 3.36 MMBD in 1970, 1.86 MMBD in 1980, 1.42 MMBD in 1990, 1.48 MMBD in 2000, and 1.61 MMBD in 2004.
- Iran produced 3.85 MMBD in 1970, 1.48 MMBD in 1980, 3.27 MMBD in 1990, 3.82 MMBD in 2000, and 4.08 MMBD in 2004.
- Saudi Arabia produced 3.85 MMBD in 1970, 10.27 MMBD in 1980, 7.11 MMBD in 1990, 9.51 MMBD in 2000, and 10.58 MMBD in 2004.

Other countries, however, show the opposite trend. Tunisia, for example, produced 0.09 MMBD in 1970, 0.12 MMBD in 1980, 0.10 MMBD in 1990, 0.08 MMBD in 2000, and 0.07 MMBD in 2004. These trends tell important stories, especially of how well the MENA countries deal with the fluctuation in the global energy demand. It also tells the story of how OPEC managed the oil production level for its MENA countries. In addition, we can see clearly how much the Gulf States dominated the production level for the MENA region and hence OPEC and the world.

Despite the importance of these stories, it is present and future production capacities that make a difference in the global energy balance. More specifically, it is the sustainable production capacity of these countries that has the largest influence on the global energy market and the international economy.

EIA Projections of Future Increases in MENA Oil Production Capacity

Virtually all major modelers project an increase in the oil production capacity of the MENA countries. As mentioned earlier, with the projected increase in global demand for oil and other sources of energy, forecasters expect that the world demand will be mostly met by increases in Gulf-OPEC production.

The ability of countries in the region to meet these requirements is, of course, sensitive to oil prices and the flow of investment capital. More generally, it is critical to any debate over whether global oil production has begun to peak and whether increases in oil supply can meet major increases in demand and keep prices relatively moderate.

The EIA estimates of MENA's oil production capacity vary sharply according to the assumption made about the price of oil: $21/barrel for the low-price case, $35/barrel for the reference case, and $48/barrel for the high-price case. In 2004–2005, the MENA oil production capacity was approximately 29.0 MMBD, and the EIA 2005 estimates called for massive increases for the MENA region's future production capacity. The following show the various estimates and the projected percentage increases:[14]

- For the low-price case, the EIA estimates the MENA region's oil production capacity to be 39.4 MMBD in 2010 (36-percent increase), 45.5 MMBD in 2015 (57-percent increase), 52.5 MMBD in 2020 (81.8-percent increase), and 59.4 MMBD in 2025 (105-percent increase).

- For the reference case, the EIA estimates the MENA region's oil production capacity to be 34.6 MMBD in 2010 (19-percent increase), 37.6 MMBD in 2015 (30-percent increase), 42.7 MMBD in 2020 (47-percent increase), and 47.8 MMBD in 2025 (65-percent increase).

- For the high-price case, it estimates the MENA region's oil production capacity to be 30.4 MMBD in 2010 (5-percent increase), 30.7 MMBD in 2015 (6-percent increase), 33.0 MMBD in 2020 (14-percent increase), and 35.2 MMBD in 2025 (21-percent increase).

These figures show that the EIA projections call for large increases in OPEC production capacity and, more specifically, in Gulf and Saudi production capacity. In its reference-case forecast, the EIA calls for the OPEC states of the Gulf alone to increase their production capacity from 20.7 MMBD in 2002, to 28.3 MMBD in 2010, 30.8 MMBD in 2015, 35.2 MMBD in 2020, and 39.3 MMBD in 2025.[15] This would mean that Gulf OPEC oil production capacity would increase from 27 percent of total world capacity in 1990 and 26 percent of world capacity in 2001, to 26–32 percent of world capacity in 2010, 24–33 percent in 2015, 24–35 percent in 2020, and 24–36 percent of world capacity in 2025 depending on the price of oil.[16] These figures would be even higher if other non-OPEC "Gulf" oil producers like Oman and Yemen were included, but it is important to note that the range does not change drastically in this forecast.

While the Gulf dominates the EIA estimate of the increase in MENA oil production capacity, the EIA estimate also projects significant increases in oil production capacity in North Africa. Algeria and Libya are estimated to increase their production from 3.2 MMBD in 2002, 3.6–4.4 MMBD in 2010, 3.6–5.2 MMBD in 2015, 4.0–6.4 MMBD in 2020, and 4.4–7.6 MMBD in 2025. The North African share of the total world production capacity, however, will not change. It will remain between 4 and 5 percent of the total.[17]

Given recent developments in the global energy market, many experts believe that these EIA projections are based on unrealistically low oil prices. Nevertheless, they provide a valuable look at what the future might be if prices stayed moderate, and there were no drastic shifts in government policies, new resources, and security risks. The following figures show these EIA projections for the period between 1990 and 2025:

- Figure 1.10 compares the EIA estimates of the regional oil production capacity between 1990 and 2025. The forecasts are based on its 2005 estimates, which assumed oil prices to be $21/barrel for the low-price case, $35/barrel for the reference case, and $48/barrel for the high-price case. In all cases, OPEC members, and more specifically the Gulf-OPEC States, are projected to dominate the world oil production capacity.

- Figure 1.11 shows the EIA's 2005 projection of increases in Middle Eastern oil production capacity by country, illustrating the critical importance of Saudi Arabia, and the role of Iran, Iraq, Kuwait, and the United Arab Emirates. Relative to world

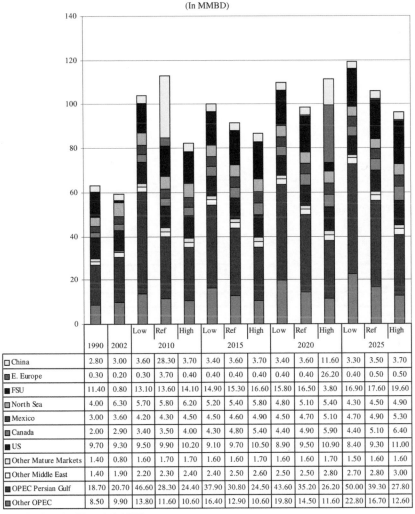

(In MMBD)

	1990	2002	2010 Low	2010 Ref	2010 High	2015 Low	2015 Ref	2015 High	2020 Low	2020 Ref	2020 High	2025 Low	2025 Ref	2025 High
☐ China	2.80	3.00	3.60	28.30	3.70	3.40	3.60	3.70	3.40	3.60	11.60	3.30	3.50	3.70
▦ E. Europe	0.30	0.20	0.30	3.70	0.40	0.40	0.40	0.40	0.40	0.40	26.20	0.40	0.50	0.50
■ FSU	11.40	0.80	13.10	13.60	14.10	14.90	15.30	16.60	15.80	16.50	3.80	16.90	17.60	19.60
▨ North Sea	4.00	6.30	5.70	5.80	6.20	5.20	5.40	5.80	4.80	5.10	5.40	4.30	4.50	4.90
■ Mexico	3.00	3.60	4.20	4.30	4.50	4.50	4.60	4.90	4.50	4.70	5.10	4.70	4.90	5.30
▨ Canada	2.00	2.90	3.40	3.50	4.00	4.30	4.80	5.40	4.40	4.90	5.90	4.40	5.10	6.40
■ US	9.70	9.30	9.50	9.90	10.20	9.10	9.70	10.50	8.90	9.50	10.90	8.40	9.30	11.00
☐ Other Mature Markets	1.40	0.80	1.60	1.70	1.70	1.60	1.60	1.70	1.60	1.60	1.70	1.50	1.60	1.60
☐ Other Middle East	1.40	1.90	2.20	2.30	2.40	2.40	2.50	2.60	2.50	2.50	2.80	2.70	2.80	3.00
■ OPEC Persian Gulf	18.70	20.70	46.60	28.30	24.40	37.90	30.80	24.50	43.60	35.20	26.20	50.00	39.30	27.80
▨ Other OPEC	8.50	9.90	13.80	11.60	10.60	16.40	12.90	10.60	19.80	14.50	11.60	22.80	16.70	12.60

Source: EIA, International Energy Outlook 2005.

Figure 1.10   EIA Projection of World Production Capacity by Region: 1990–2025

production, the MENA region is estimated to play a major role in determining the world oil supply for the next 20 years.

• Figure 1.12 illustrates the impact and importance of high oil prices on the estimated rate of increase in Middle Eastern oil production capacity, illustrating that price is one of the key uncertainties. It also highlights sensitivity and elasticity analysis of the EIA estimates of MENA's supply trends.

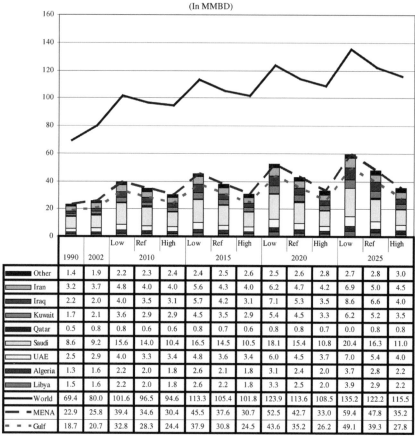

(In MMBD)

	1990	2002	Low 2010	Ref 2010	High 2010	Low 2015	Ref 2015	High 2015	Low 2020	Ref 2020	High 2020	Low 2025	Ref 2025	High 2025
Other	1.4	1.9	2.2	2.3	2.4	2.4	2.5	2.6	2.5	2.6	2.8	2.7	2.8	3.0
Iran	3.2	3.7	4.8	4.0	4.0	5.6	4.3	4.0	6.2	4.7	4.2	6.9	5.0	4.5
Iraq	2.2	2.0	4.0	3.5	3.1	5.7	4.2	3.1	7.1	5.3	3.5	8.6	6.6	4.0
Kuwait	1.7	2.1	3.6	2.9	2.9	4.5	3.5	2.9	5.4	4.5	3.3	6.2	5.2	3.5
Qatar	0.5	0.8	0.8	0.6	0.6	0.8	0.7	0.6	0.8	0.8	0.7	0.0	0.8	0.8
Saudi	8.6	9.2	15.6	14.0	10.4	16.5	14.5	10.5	18.1	15.4	10.8	20.4	16.3	11.0
UAE	2.5	2.9	4.0	3.3	3.4	4.8	3.6	3.4	6.0	4.5	3.7	7.0	5.4	4.0
Algeria	1.3	1.6	2.2	2.0	1.8	2.6	2.1	1.8	3.1	2.4	2.0	3.7	2.8	2.2
Libya	1.5	1.6	2.2	2.0	1.8	2.6	2.2	1.8	3.3	2.5	2.0	3.9	2.9	2.2
World	69.4	80.0	101.6	96.5	94.6	113.3	105.4	101.8	123.9	113.6	108.5	135.2	122.2	115.5
MENA	22.9	25.8	39.4	34.6	30.4	45.5	37.6	30.7	52.5	42.7	33.0	59.4	47.8	35.2
Gulf	18.7	20.7	32.8	28.3	24.4	37.9	30.8	24.5	43.6	35.2	26.2	49.1	39.3	27.8

Source: EIA, International Energy Outlook, various editions.

Figure 1.11   EIA Projection of MENA Oil Production Capacity by Country: 1990–2025

IEA Projections of Future Increases in MENA Oil Production Capacity

The IEA makes generally similar projections, although it uses different time periods and definitions of the regions to be assessed. The IEA, in its 2005 reference scenario, estimated that the MENA oil production capacity is 29.0 MMBD in 2004, and will rise to 33.0 MMBD in 2010 (13-percent increase), 41.8 MMBD in 2020 (44-percent increase), and 50.5 MMBD in 2030 (74-percent increase).[18]

Put differently, the MENA oil production capacity will increase by 74 percent during the forecast period. Most of this increase, as Figure 1.13 shows, will come from the Gulf-OPEC members. The Gulf production capacity is estimated to increase from 20.0 MMBD in 2004 to 23.6 MMBD in 2010 (14-percent increase),

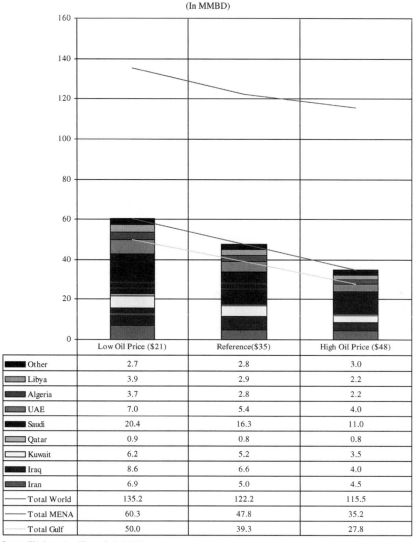

(In MMBD)

	Low Oil Price ($21)	Reference($35)	High Oil Price ($48)
Other	2.7	2.8	3.0
Libya	3.9	2.9	2.2
Algeria	3.7	2.8	2.2
UAE	7.0	5.4	4.0
Saudi	20.4	16.3	11.0
Qatar	0.9	0.8	0.8
Kuwait	6.2	5.2	3.5
Iraq	8.6	6.6	4.0
Iran	6.9	5.0	4.5
Total World	135.2	122.2	115.5
Total MENA	60.3	47.8	35.2
Total Gulf	50.0	39.3	27.8

Sources: EIA, International Energy Outlook 2005.

Figure 1.12   Variations in EIA Estimate of MENA Oil Production Capacity in 2025 by Prices

to 29.9 MMBD in 2020 (44-percent increase), to 36.2 MMBD in 2030 (75-percent increase).[19]

The estimated increases are projected to come from a variety of sources, but not all the countries are expected to increase through 2030. Algeria's production capacity is projected to increase then decline during this period from 1.9 MMBD in 2004, to 2.2 MMBD in 2010, to 1.9 MMBD in 2020, to 1.6 MMBD in 2030. Egypt's

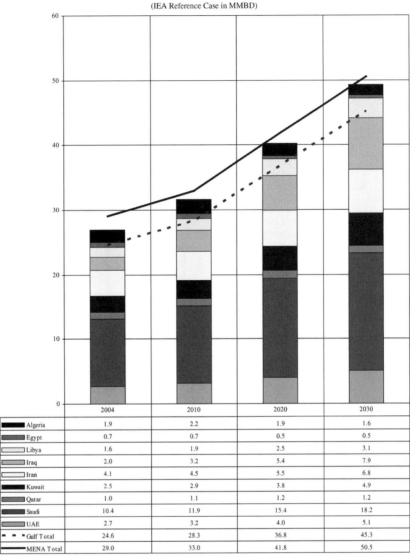

(IEA Reference Case in MMBD)

	2004	2010	2020	2030
Algeria	1.9	2.2	1.9	1.6
Egypt	0.7	0.7	0.5	0.5
Libya	1.6	1.9	2.5	3.1
Iraq	2.0	3.2	5.4	7.9
Iran	4.1	4.5	5.5	6.8
Kuwait	2.5	2.9	3.8	4.9
Qatar	1.0	1.1	1.2	1.2
Saudi	10.4	11.9	15.4	18.2
UAE	2.7	3.2	4.0	5.1
Gulf Total	24.6	28.3	36.8	45.3
MENA Total	29.0	33.0	41.8	50.5

Source: IEA, World Energy Outlook 2005.

Figure 1.13   IEA Projection of Oil Production Capacity by Country: 2004–2030

production capacity is also expected to decline from 0.7 MMBD in 2004, to 0.7 MMBD in 2010, to 0.5 MMBD in 2020, to 0.5 MMBD in 2030.

The two most critical countries in terms of the IEA's projected increases are Saudi Arabia and Iraq. In the case of Saudi Arabia, the IEA estimates that the Kingdom's production capacity will increase from 10.4 MMBD in 2004, to 11.9 MMBD in 2010, top 15.4 MMBD in 2020, to 18.2 MMBD in 2025.[20] These estimates also track with those of the EIA 2005 reference-case estimates, which projected that the Kingdom's oil production capacity will reach 14.0 MMBD in 2010, 14.5 MMBD in 2015, 15.4 MMBD in 2020, and 16.3 MMBD in 2025.

It is important to note, however, that many current and former oil officials in Saudi Arabia believe that these forecasts are too high, and, short of meaningful technological breakthroughs, the Kingdom will not be able to meet the 18.2 MMBD by 2030. Most oil company experts also feel the IEA and EIA estimates reflect what global markets want rather than what Saudi Arabia plans to provide or will find economically and technically feasible.

In the case of Iraq, it is even harder to make sensible projections because the current data are even less accurate. Any forecast must take into account the former regime's mismanagement of its oil infrastructure, the impact of UN sanctions on Iraq's oil industry, ongoing security risks, the politicization of the oil industry, and the wide range of uncertainties surrounding Iraq's future. Many Iraqi oil experts believe that the ongoing insurgency and the slow economic progress have pushed Iraq's oil development back several years.

Even so, the IEA estimates that Iraq's oil production capacity will increase from 2.0 MMBD in 2004, to 3.2 MMBD in 2010, to 5.4 MMBD in 2020, to 7.9 MMBD in 2030.[21] These are similar to the EIA estimates. In its reference-case scenario, the EIA estimates that Iraq's oil production will increase to 3.5 MMBD in 2010, 4.2 MMBD in 2015, 5.3 MMBD in 2020, and 6.6 MMBD in 2025.[22]

Current and Future MENA Oil Exports

Estimates of past, current, and projected trends in oil exports follow different patterns from increases in oil production capacity for several reasons. Oil prices and OPEC quotas have a major impact on what countries actually produce, and producers consume significant and growing portions of their domestic production. Most sources show that the MENA region, however, will retain massive surplus production capacity relative to domestic demand. This explains why its share of world exports has been, and is projected to be, much higher than its share of total production or production capacity.

Different Projections of Export Trends

According to estimates in the BP *Statistical Review of World Energy*, the Middle East produced an annual average of 21.1 MMBD in 2002, 23.1 MMBD in 2003, and 24.5 MMBD in 2004, which represented approximately 30 percent of the world's total production of 80.2 MMBD. In 2004, the Middle East also exported an average of 19.63 MMBD, or 41 percent of the world's total of 48.2 MMBD. This is up from its 2002 level of 18.1 MMBD.[23]

If the four oil exporters in North Africa—Egypt, Algeria, Libya, and Tunisia—are added to this total to create a figure for the MENA region, this adds average annual exports of 2.91 MMBD or 6 percent of the world's total oil exports. This brings the total Middle East and North African region exports to 22.5 MMBD or 46.7 percent of the world's total oil exports.[24]

The IEA estimates the MENA region exported nearly 22.3 MMBD of oil in 2004 and that the region's export capacity will steadily increase over the next 25 years. Its *World Energy Outlook 2005* projects that the MENA region will export 25.0 MMBD (or a 13-percent increase from 2004) in 2010, 31.8 MMBD (42-percent increase) in 2020, and 38.7 MMBD (73-percent increase) in 2030.[25]

Figures 1.14 and 1.15 show the EIA estimates of the projected increases in the global dependence on Gulf oil exports relative to other oil-producing regions. The Gulf-OPEC States exported an average of 16.9 MMBD, or 30 percent of the world total of 56.3 MMBD in 2001, 16.7 MMBD, or 30 percent of the world total of 55.5 MMBD in 2002, and is estimated to export 35.4 MMBD, or 41 percent of the world total of 86.0 MMBD in 2025. If one includes the North African states Algeria and Libya, the exports climb to 19.5 MMBD, or 35 percent in 2002. The EIA also projects that the whole MENA region's exports will reach 40.7 MMBD, or 47 percent, in 2025.[26]

While estimates of future export trends are no more certain than estimates of oil production capacity, it is again clear that it would take a massive breakthrough in technology or discoveries of reserves outside the Middle East to change these trends. These totals also understate the true importance of the MENA region because the EIA, unlike BP and the IEA, does not issue an estimate for the entire Middle East or MENA region as distinguished from the Gulf, and OPEC country estimates for the region exclude exports from Oman, Yemen, and the Levant.

The Direction of MENA Oil Exports

Under most conditions, the normal day-to-day destination of MENA oil exports is strategically irrelevant. Oil is a global commodity, which is distributed to meet the needs of a global market based on process bid by importers acting in global competition. With the exception of differences in price because of crude type and transportation costs, all buyers compete equally for the global supply of available exports, and the direction and flow of exports changes according to marginal price relative to demand.

As a result, the percentage of oil that flows from the MENA region to the United States under normal market conditions has little strategic or economic importance. If a crisis occurs, or drastic changes take place in prices, the United States will have to pay the same globally determined price as any other nation, and the source of U.S. imports will change accordingly. Moreover, the United States is required to share all imports with other Organisation for Economic Co-operation Development (OECD) countries in a crisis under the monitoring of the International Energy Agency.

Most estimates do, however, project that China will dominate the growth of consumption for the next two decades—followed by India, other East Asian states, the

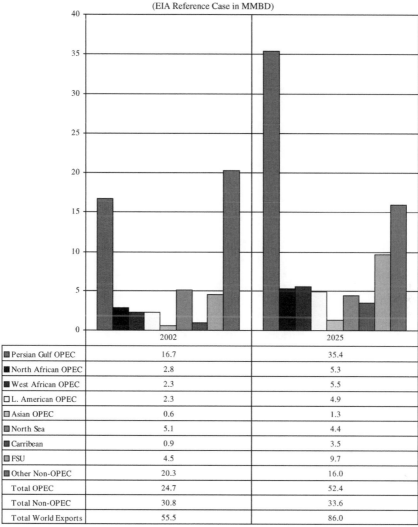

(EIA Reference Case in MMBD)

	2002	2025
■ Persian Gulf OPEC	16.7	35.4
■ North African OPEC	2.8	5.3
■ West African OPEC	2.3	5.5
□ L. American OPEC	2.3	4.9
■ Asian OPEC	0.6	1.3
■ North Sea	5.1	4.4
■ Carribean	0.9	3.5
■ FSU	4.5	9.7
■ Other Non-OPEC	20.3	16.0
Total OPEC	24.7	52.4
Total Non-OPEC	30.8	33.6
Total World Exports	55.5	86.0

Source: EIA, International Energy Outlook 2005.

Figure 1.14 The Rising Importance of Gulf Exports Relative to Other Exports in Meeting World Demand: 2002 vs. 2025

Middle East states, Africa, and the United States. Japanese oil consumption, on the other hand, is estimated to plateau. They also project that these shifts will occur at a time when the United States and other industrialized states are increasingly dependent on both the health of the global economy and increasing flows of oil exports to major suppliers outside the "industrialized world." With the exception of Latin America, Mexico, and Canada, all of America's major trading partners are

Figure 1.15   EIA Estimate of World Oil Exports by Supplier and Destination: 2002–2025 (in MMBD)

Exporting Regions	North America	Western Europe	Asia	Importing Regions Total Mature	Pacific	China	Rest of World	Total Emerging
				2002				
OPEC	6.2	5.6	4.8	16.6	4.3	0.9	2.9	8.1
Persian Gulf	2.8	2.9	4.4	10.1	3.2	0.9	2.5	6.6
North Africa	0.6	2.1	0.0	2.7	0.1	0.0	0.0	0.1
West Africa	1.1	0.5	0.1	1.7	0.5	0.0	0.1	0.6
South America	1.7	0.1	0.1	1.9	0.1	0.0	0.3	0.4
Asia	0.0	0.0	0.2	0.2	0.4	0.0	0.0	0.4
Non-OPEC	7.0	11.8	1.5	20.3	3.3	1.3	5.9	10.5
North Sea	0.6	4.5	0.0	5.1	0.0	0.0	0.0	0.0
Caribbean Basin	0.6	0.1	0.0	0.7	0.1	0	0.2	0.2
FSU	0.2	3.6	0.3	4.2	0.2	0.0	0.3	0.3
Other Non-OPEC	5.5	3.6	1.2	10.3	3	1.3	5.7	10
World Total	**13.2**	**17.4**	**6.3**	**36.9**	**7.6**	**2.2**	**8.8**	**18.6**
				2025				
OPEC	11.9	8.8	6.2	26.9	12	7.3	6.2	25.5
Persian Gulf	5.8	4.5	5.1	15.4	8.7	6.4	4.9	20
North Africa	0.5	3.1	0.1	3.7	0.8	0.3	0.5	1.6
West Africa	1.6	1.1	0.3	3	1.8	0.5	0.2	2.5
South America	3.9	0.1	0.4	4.4	0.1	0	0.4	0.5
Asia	0.1	0	0.3	0.4	1.6	0.1	0.2	0.9
Non-OPEC	9.2	10.1	1.2	20.5	4.7	3.4	5.0	13.1
North Sea	0.5	3.4	0.0	3.9	0.3	0.0	0.2	0.5
Caribbean Basin	1.4	0.5	0.2	2.1	0.6	0.0	0.8	1.4
FSU	0.5	3.3	0.6	4.4	0.7	3.1	1.5	5.3
Other Non-OPEC	6.8	2.9	0.4	10.1	3.1	0.3	2.5	5.9
World Total	**21.1**	**18.9**	**7.4**	**47.4**	**16.7**	**10.7**	**11.2**	**38.6**

Source: EIA, *International Energy Outlook 2005*, p. 34.

critically dependent on Middle Eastern oil exports and are projected to become more dependent in the future.

Dependence on Indirect Imports

The size of direct imports of petroleum is only a partial measure of strategic dependence. The U.S. economy is dependent on energy-intensive imports from Asia and other regions, and what comes around must literally go around. While the EIA and the IEA do not make estimates of indirect imports of Middle Eastern oil in terms of the energy required to produce the finished goods, because the United States imports them from countries that are dependent on Middle Eastern exports, analysts guess that they would add at least 1 MMBD to total U.S. oil imports.

To put this figure in perspective, direct U.S. oil imports increased from an annual average of 7.9 MMBD in 1992 to 11.3 MMBD in 2002, and 2.6 MMBD worth of U.S. petroleum imports came directly from the Middle East in 2002. In 2004, according to BP, the U.S. total imports totaled 12.8 MMBD of which 2.97 MMBD came from the MENA region, or 23.2 percent.[27] If indirect U.S. imports in the form of manufactured goods dependent on imports of Middle Eastern oil were included, the resulting figure might well be 30–40 percent higher than the figure for direct imports.

In any case, the failure of the EIA and the IEA to explicitly model such indirect imports, and their steady growth, is a long-standing and critical failure in energy analysis and policy. It seems almost certain that the future increase in such indirect imports will, for example, vastly exceed any benefits in increased domestic energy supply that will result from the energy bill that was passed by the U.S. Congress in the summer of 2005.

Dependence on the Flow of Oil to the Global Economy

The United States and other industrialized states are increasingly dependent on the health of the global economy. U.S. economic activity and growth are dependent on how well the economies of Europe, Asia, and Latin America function. With the exception of Latin America, Mexico, and Canada, all of America's major trading partners are critically dependent on Middle Eastern oil exports.

Figure 1.16 shows how the Middle East and North Africa dominated interarea movements of petroleum literally "fuel" the exports of Asia to the rest of the world and are critical sources of indirect energy exports to other regions. In 2004, the Middle East and North Africa supplied 5.1 MMBD of 12.5 MMBD of European imports (41 percent). MENA exporters supplied 4.2 MMBD of Japanese imports of 5.2 MMBD (80.8 percent). MENA countries supplied 1.3 MMBD of China's imports of 3.4 MMBD (38.3 percent, and growing steadily in recent years), 0.13 MMBD of Australasia's imports of 0.69 MMBD (19.5 percent), and 7.3 MMBD of some 9.2 MMBD in imports by other Asian and Pacific states (78.9 percent).[28]

The EIA projected that the global economy will also grow far more dependent on the Middle Eastern and North African oil imports in the future. The EIA's *International Energy Outlook 2005* projects North American imports of MENA oil to increase from 3.4 MMBD in 2002 to 6.2 MMBD in 2025—an increase of 91 percent, almost all of which will go to the United States. Exports to western Europe are estimated to increase from 5.0 MMBD to 7.6 MMBD, or by 62 percent.

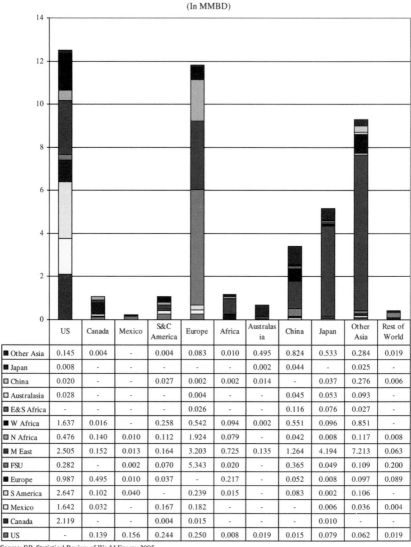

	US	Canada	Mexico	S&C America	Europe	Africa	Australasia	China	Japan	Other Asia	Rest of World
■ Other Asia	0.145	0.004	-	0.004	0.083	0.010	0.495	0.824	0.533	0.284	0.019
■ Japan	0.008	-	-	-	-	-	0.002	0.044	-	0.025	-
▢ China	0.020	-	-	0.027	0.002	0.002	0.014	-	0.037	0.276	0.006
▢ Australasia	0.028	-	-	-	0.004	-	-	0.045	0.053	0.093	-
■ E&S Africa	-	-	-	-	0.026	-	-	0.116	0.076	0.027	-
■ W Africa	1.637	0.016	-	0.258	0.542	0.094	0.002	0.551	0.096	0.851	-
▢ N Africa	0.476	0.140	0.010	0.112	1.924	0.079	-	0.042	0.008	0.117	0.008
■ M East	2.505	0.152	0.013	0.164	3.203	0.725	0.135	1.264	4.194	7.213	0.063
▣ FSU	0.282	-	0.002	0.070	5.343	0.020	-	0.365	0.049	0.109	0.200
■ Europe	0.987	0.495	0.010	0.037	-	0.217	-	0.052	0.008	0.097	0.089
▢ S America	2.647	0.102	0.040	-	0.239	0.015	-	0.083	0.002	0.106	-
▢ Mexico	1.642	0.032	-	0.167	0.182	-	-	-	0.006	0.036	0.004
■ Canada	2.119	-	-	0.004	0.015	-	-	-	0.010	-	-
▣ US	-	0.139	0.156	0.244	0.250	0.008	0.019	0.015	0.079	0.062	0.019

Source: BP, Statistical Review of World Energy 2005.

Figure 1.16   BP Estimates of Interarea Movements: 2004

These projections are based on several important assumptions: (a) major increases will take place in oil exports from the former Soviet Union (FSU), as well as the MENA region; (b) conservation will limit the scale of European imports from the Middle East; (c) industrialized Asia—driven by Japan—will increase its imports from 4.4 MMBD in 2002 to 5.6 MMBD in 2025, or nearly 27 percent; (d) China will increase its imports from 0.9 MMBD to 6.7 MMBD, or by nearly 644 percent;

and (e) Pacific Rim states will increase imports from 3.3 MMBD to 9.5 MMBD, or by 187 percent.[29]

The trends projected by the IEA in 2002 are similar to the trends projected by the EIA. The IEA projects that total interregional trade in oil will increase from 32 MMBD in 2000 to 42 MMBD in 2010 and 66.1 MMBD in 2030, and Middle Eastern exports (less North Africa) will increase from 19 MMBD in 2000 to 46 MMBD in 2030. Most of these additional exports will go to Asia, with China emerging as the largest market, followed by India. The rise in U.S. imports will be limited by increased exports from Canada, because of production from tar sands, from Mexico, and from Sub-Saharan Africa.[30]

Figure 1.17 shows the breakdown of the MENA region net trade or export capacity by country. These IEA estimates show an increase in MENA export capacity between 2004 and 2030. In 2004, the MENA region is estimated to have traded 22.3 MMBD of oil. The IEA projected in 2005 that the MENA region will export 25.0 MMBD in 2010 (12-percent increase), 31.8 MMBD in 2020 (42-percent increase), and 38.7 MMBD in 2030 (73.5-percent increase).[31]

General Patterns of Global Oil Dependence

If one looks at the overall patterns in Asian demand, the EIA estimate for 2005 projected the trends shown in Figure 1.15. It is important to note that the projected total growth for "other Asia" nearly totals the growth in China. Total Asian demand is estimated to rise from 21.5 MMBD in 2002 to 35.6–45.50 MMBD in 2025. India's oil demand is estimated to increase from 2.20 MMBD in 2002 to 4.30–5.40 MMBD in 2025. Other Asia's oil demand is estimated to rise from 5.6 MMBD in 2002 to 9.80–13.4 MMBD in 2025. South Korea's demand for oil is estimated to rise from 2.20 MMBD in 2002 to 2.60–3.40 MMBD in 2025.[32]

According to China's state media reports, China imported 79.9 million tons of oil in the first three quarters of 2004, which represented a 40-percent increase from the first eight months of 2003.[33] In 2002, China consumed 5.0 MMBD. According to EIA 2005 high-price estimates, this number could triple by 2025 (12.50 MMBD for the low-price case, 14.50 MMBD for the reference case, and 16.1 MMBD for the high-price case).[34]

According to the BP *Statistical Review of World Energy 2005*, Chinese imports totaled 3.40 MMBD in 2004. China imported 0.15 MMBD from the United States, 0.038 MMBD from South and Central America, 0.052 MMBD from Europe, 0.365 MMBD from the FSU, 1.264 MMBD from the Middle East, 0.709 MMBD from Africa, 0.045 MMBD from Australasia, 0.044 MMBD from Japan, 0.824 from other Asia Pacific nations, and 0.010 MMBD from others.[35]

By the end of 2005, China consumed approximately 6.4 MMBD—second to U.S. consumption of 21.0 MMBD. China's consumption by 2020 is projected to triple. The growth of Chinese oil demand is higher than its domestic supply. China's domestic production could reach 3.8 MMBD in 2020, but its demand is likely to be more than three times as high. In addition, dependence on Middle Eastern oil

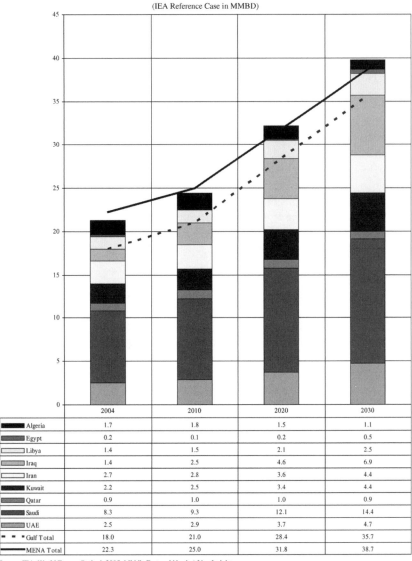

(IEA Reference Case in MMBD)

	2004	2010	2020	2030
Algeria	1.7	1.8	1.5	1.1
Egypt	0.2	0.1	0.2	0.5
Libya	1.4	1.5	2.1	2.5
Iraq	1.4	2.5	4.6	6.9
Iran	2.7	2.8	3.6	4.4
Kuwait	2.2	2.5	3.4	4.4
Qatar	0.9	1.0	1.0	0.9
Saudi	8.3	9.3	12.1	14.4
UAE	2.5	2.9	3.7	4.7
Gulf Total	18.0	21.0	28.4	35.7
MENA Total	22.3	25.0	31.8	38.7

Source: IEA, World Energy Outlook 2005, Middle East and North Africa Insights.

Figure 1.17   IEA Projection of Oil Net Trade by Country: 2004–2030

has increased from 39.79 percent of its imports in 1994, to 50.99 percent in 2002, and to over 50 percent in 2004.[36]

If one looks at western Europe, oil consumption is estimated to rise slowly from 13.8 MMBD in 2002 to 13.7–14.4 MMBD in 2010, 13.8–14.8 MMBD in 2015, 13.8–15.2 MMBD in 2020, and 14.1–15.9 MMBD in 2025. Eastern European growth will also be somewhat faster but much lower. Eastern European countries consumed 1.4 MMBD in 2002. It is estimated to rise to 1.6 MMBD in 2010, 1.7–1.9 MMBD in 2015, 1.8–2.1 MMBD in 2020, and 2.8–2.4 MMBD in 2025.[37]

To put these figures in perspective, total European oil use will rise from 15.2 MMBD in 2002 to 18.2 MMBD in 2025. This is an estimated growth of 0.8 MMBD to 3.0 MMBD. If one looks at the projected growth for the United States over the same period, it is an increase of 2.1 MMBD to 6.2 MMBD by 2025. Russia is estimated to increase oil consumption from 2.6 MMBD in 2002 to 3.1–3.9 MMBD in 2025 and the rest of FSU from 1.5 MMBD in 2002 to 1.8–2.3 MMBD in 2025.[38]

The Trends in U.S. Petroleum Imports

U.S. oil imports are only a subset of U.S. strategic dependence on Middle East oil exports. As has been noted earlier, the United States is dependent on the overall health of the global economy and on large amounts of indirect energy imports in the form of manufactured goods from Asia and other nations dependent on Middle East oil. The United States must also compete for the global supply of oil exports on market terms in any short-term crisis, or longer-term shortfall in MENA exports, and it is the global supply of oil exports relative to global demand, not where the United States gets oil at any given time, that determines availability and price to the United States as well as all other importing nations.

These realities are reflected in the past patterns of U.S. dependence on oil imports from the Middle East. The EIA reports wide fluctuations in U.S. oil imports over time. If one looks only at total U.S. imports of petroleum, imports from all sources reached 6.2 MMBD in 1973. Since then, it steadily increased to 6.9 MMBD in 1980, 8.0 MMBD in 1990, 11.4 MMBD in 2000, 13.1 MMBD in 2004, and 13.4 MMBD in 2005.[39]

Figure 1.18 shows recent trends in U.S. import dependence on Gulf and other OPEC oil. The United States is dependent on the Middle East for only part of its imports, and there have been no consistent trends in the percentage of imports the United States gets from OPEC and the Gulf, but it is all too clear that U.S. oil imports are increasing.

If one looks at OPEC exports as a percent of U.S. imports, these ranged from 47.8 percent in 1973, and 51.9 percent MMBD in 1992, to 39.9 percent MMBD in 2002, and 43.6 percent MMBD in 2004. If one looks at Gulf exports as a percent of U.S. imports, they ranged from 13.6 percent in 1973, and 22.5 percent MMBD in 1992, to 19.7 percent MMBD in 2002, and 19.3 percent MMBD in 2004.

The United States has become progressively more dependent on both a growing volume of imports and steadily growing imports from troubled countries and regions. Direct U.S. petroleum imports increased from an annual average of 6.3 MMBD in 1973, to 7.9 MMBD in 1992, to 11.3 MMBD in 2002, and to 12.9 MMBD in 2004. Some 2.5 MMBD worth of U.S. petroleum imports came directly from the Middle East in 2004.[40] Additionally, the average U.S. petroleum imports from the Gulf alone equaled approximately 2.3 MMBD in 2005,

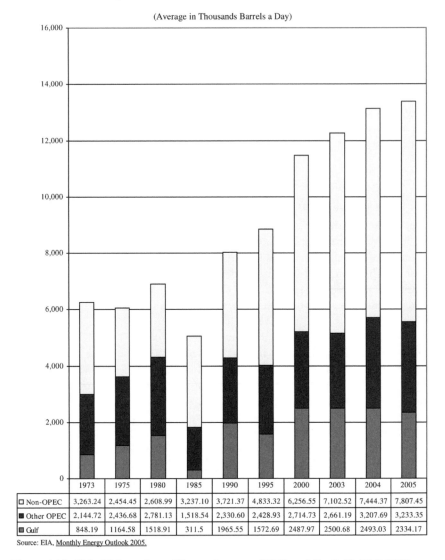

Figure 1.18   U.S. Oil Imports and Dependence on OPEC and the Gulf: 1973–2005

2.4 MMBD in 2004, 2.5 MMBD in 2003, 2.2 MMBD in 2002, 2.7 MMBD in 2001, and 2.4 MMBD in 2000.[41]

Looking toward the future, the EIA forecast in the *Annual Energy Outlook 2005* that total U.S. petroleum imports would reach 20.0 MMBD by 2025, as Figure 1.19 shows. This projection was based on somewhat more realistic oil prices (low-price case, $20.99; reference case, $30.31; high-price case, $39.24) than those used

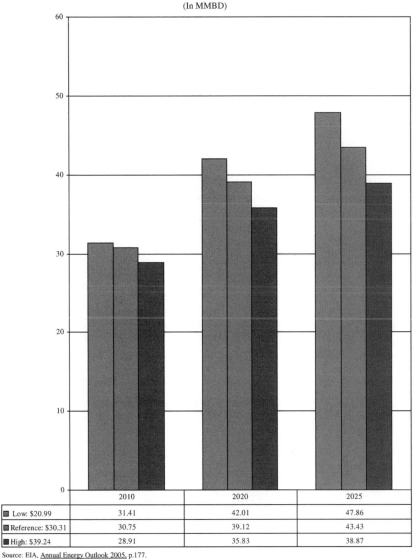

(In MMBD)

	2010	2020	2025
◩ Low: $20.99	31.41	42.01	47.86
◪ Reference: $30.31	30.75	39.12	43.43
◼ High: $39.24	28.91	35.83	38.87

Source: EIA, Annual Energy Outlook 2005, p.177.

Figure 1.19   Core Petroleum U.S. Import Elasticity—Forecast for 2025

in the *International Energy Outlook 2004*, which have been the basis for the previous discussion of global trends.

The EIA describes the growing U.S. need for imports as follows:[42]

Total U.S. gross petroleum imports are projected to increase in the reference case from 12.3 million barrels per day in 2003 to 20.2 million in 2025 [Figure 1.20]. Crude oil accounts for most of the increase in imports, because distillation capacity at U.S.

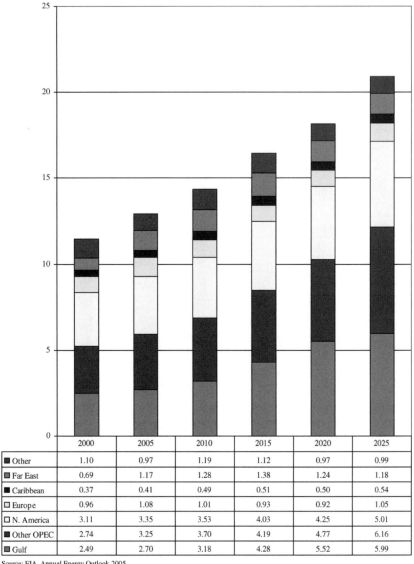

	2000	2005	2010	2015	2020	2025
■ Other	1.10	0.97	1.19	1.12	0.97	0.99
▨ Far East	0.69	1.17	1.28	1.38	1.24	1.18
■ Caribbean	0.37	0.41	0.49	0.51	0.50	0.54
□ Europe	0.96	1.08	1.01	0.93	0.92	1.05
□ N. America	3.11	3.35	3.53	4.03	4.25	5.01
■ Other OPEC	2.74	3.25	3.70	4.19	4.77	6.16
▨ Gulf	2.49	2.70	3.18	4.28	5.52	5.99

Source: EIA, Annual Energy Outlook 2005.

Figure 1.20   U.S. Gross Petroleum Imports by Source, 2000–2025

refineries is expected to be more than 5.5 million barrels per day higher in 2025 than it was in 2003. Gross imports of refined petroleum, including refined products, unfinished oils, and blending components, are expected to increase by almost 60 percent from 2003 to 2025.

Crude oil imports from the North Sea are projected to decline gradually as North Sea production ebbs. Significant imports of petroleum from Canada and Mexico are expected to continue, with much of the Canadian contribution coming from the development of its enormous oil sands resource base. West Coast refiners are expected to import small volumes of crude oil from the Far East to replace the declining production of Alaskan crude oil. The Persian Gulf share of total gross petroleum imports, 20.4 percent in 2003, is expected to increase to almost 30 percent in 2025; and the OPEC share of total gross imports, which was 42.1 percent in 2003, is expected to be above 60 percent in 2025.

Most of the increase in refined product imports is projected to come from refiners in the Caribbean Basin, North Africa, and the Middle East, where refining capacity is expected to expand significantly. Vigorous growth in demand for lighter petroleum products in developing countries means that U.S. refiners are likely to import smaller volumes of light, low-sulfur crude oils.

If one looks at Figure 1.20, this estimate indicates that moderate oil prices will lead to major increases in U.S. imports from the Gulf (2.5 to 6.0 MMBD), the Americas (3.1 to 5.0 MMBD), and "other," including North Africa (2.7 to 6.2 MMBD).

Figure 1.19 shows, however, that future U.S. imports will vary sharply according to price. According to March 2005 statistics from the EIA, the oil price "collapse" of late 1997 and 1998 cut U.S. net oil import costs during 1998 by around $20 billion (to $44 billion), compared to the previous two years. Increased oil prices since then have increased U.S. net oil import costs: to $60 billion in 1999, $109 billion in 2000, $94 billion in both 2001 and 2002, and $122 billion during 2003. For the first ten months of 2004, U.S. net oil import costs were running about 31 percent higher than during the same period in 2003. Oil currently accounts for about one-fourth of the total U.S. merchandise trade deficit.

The EIA estimates that if future prices are low ($20.99/barrel), imports will rise to 47.86 MMBD in 2025. If prices are moderate ($30.31/barrel), imports are still 43.43 MMBD. If prices rise to $39.87/barrel, however, U.S. imports are only 38.87 MMBD, and they would be far lower at $50, $60, $70, or more per barrel. Even the "high-price" case leaves the United States with nearly 60-percent dependence on oil imports in 2025, but the impact of this dependence on world supply is far lower than if oil prices are low or moderate.

As noted earlier, the size of U.S. direct imports is only part of the story since the U.S. economy is highly dependent on energy-intensive products. This growing dependence does give the United States a strong strategic interest in ensuring that the global economy can draw on oil exports from as many regions as possible. As a result, national security and strategic planners in consuming nations have looked at different regions such as West Africa and the Caspian Sea to try to find other sources of supply, in addition to studying the feasibility of alternative energy technology. So far, such efforts have failed to find both the necessary oil reserves and a mix of nations that are more (or even "as") stable as those of the Gulf and North Africa.

The United States has reached out into Central Asia as part of such efforts. For example, the U.S. military is investing time and effort in emerging oil regions. The

United States is planning to spend $100 million to build up the Caspian Guard, a network of police and Special Operation forces, to protect the new Baku-Tblisi-Ceyhan pipeline, from the Caspian Sea through the Caucasus Mountains. The $100 million will also be put toward protecting other energy infrastructure to limit any supply disruption from the central Asian region. The Caspian Guard also opened a radar-equipped command center in 2003 in Baku, Azerbaijan, to monitor the oil production and export infrastructure. Most of the Caspian Sea oil is exported to Europe, but if supply is disrupted, oil prices will likely rise and that will have a direct influence on the U.S. energy security.[43]

It must be stressed, however, that such efforts result more in aiding the global economy than in securing the United States directly. With the exception of differences in price because of crude type and transportation costs, buyers and importers compete equally for the global available supply and exports. As a result, the percentage of oil that flows from the MENA region to the United States under normal market conditions has little strategic or economic importance.

Changes in the Nature of Petroleum Imports from the MENA Region

More is involved in analyzing the importance of MENA oil exports than estimating the export of crude oil. The Middle Eastern and North African states are steadily attempting to increase profit margins by producing and exporting refined oil products, rather than selling crude. At the same time, some countries—such as the United States—have created major permitting and environmental barriers to creating new refineries. As a result, the nature of Middle East exports will shift sharply from crude oil to product over the coming decades.

Middle Eastern refinery distillation capacity has already increased from 5.0 MMBD in 1990 (8 percent of world capacity) to 5.9 MMBD in 2000 (7 percent). In its *World Energy Outlook 2005*, the IEA projects that the MENA refinery capacity will increase from 8.8 MMBD in 2004, to 10.9 MMBD in 2010 (23.8-percent increase), to 13.4 MMBD in 2020 (52.2-percent increase), and to 16.0 MMBD in 2030 (81.8-percent increase). Total MENA capacity is projected to nearly double, and the same is true for the Gulf total capacity. In 2004, the Gulf refining capacity was 5.9 MMBD and is estimated to reach 11.1 MMBD in 2030.[44]

Figure 1.21 shows the IEA's country-by-country refinery distillation capacity projections. Country plans to increase refinery capacity are discussed in detail in later chapters, but this figure shows the overall trend in refining capacity for the MENA region for the next 25 years. These numbers are based on the IEA's 2005 projections.

The largest increase relative to current capacity is projected for Saudi Arabia. Its refining capacity is estimated to double from 2.1 MMBD in 2004, to 2.6 MMBD in 2010, to 3.4 MMBD in 2020, to 4.5 MMBD in 2030. The Kingdom was followed by Iran with a refining capacity of 1.5 MMBD in 2004 and is estimated to increase to 1.7 MMBD in 2010, to 2.2 MMBD in 2020, and to 2.6 MMBD in

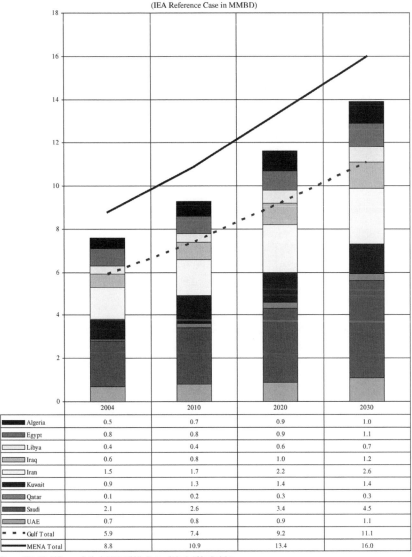

(IEA Reference Case in MMBD)

	2004	2010	2020	2030
Algeria	0.5	0.7	0.9	1.0
Egypt	0.8	0.8	0.9	1.1
Libya	0.4	0.4	0.6	0.7
Iraq	0.6	0.8	1.0	1.2
Iran	1.5	1.7	2.2	2.6
Kuwait	0.9	1.3	1.4	1.4
Qatar	0.1	0.2	0.3	0.3
Saudi	2.1	2.6	3.4	4.5
UAE	0.7	0.8	0.9	1.1
Gulf Total	5.9	7.4	9.2	11.1
MENA Total	8.8	10.9	13.4	16.0

Source: IEA, World Energy Outlook 2005, Middle East and North Africa Insights.

Figure 1.21   IEA Projection of Refining Capacity by Country: 2004–2030

2030. Iraq's refining capacity is estimated to steadily increase from 0.6 MMBD in 2004, to 0.8 MMBD in 2010, to 1.0 MMBD in 2020, and 1.2 MMBD in 2030.

In North Africa, only Algeria is estimated to see large increases in capacity, which is estimated to double during the forecast period. Algeria's refinery and product capacity are estimated to increase from 0.5 MMBD in 2004, to 0.7 MMBD in 2010, to 0.9 MMBD in 2020, and to 1.0 MMBD in 2030. This represents a

100-percent increase over the next 25 years. Egypt and Libya's refining capacity are estimated to increase much more moderately.

Such a shift to product exports does not necessarily alter importer dependence in strategic terms. It can, however, lead to greater dependence on a given Middle Eastern supplier, because a given exporter produces the products given importers need for their economy and industries. It can reduce the flexibility of global markets in substituting for Middle Eastern oil because there may be no source of similar refinery or production capacity that can provide substitutes. A shift to product exports also reduces the total volume of product shipped, although it increases its value, making MMBD a less valid measure of dependence on oil imports.

The Factors That Could Sustain High Demand for Oil

The short-term trends in demand can easily push against the limits imposed by global supply. According to the U.S. DOE, world oil demand will exceed 86.0 MMBD in the fourth quarter of 2005, which represents a 1.6–1.9 MMBD increase. Furthermore, demand is projected to increase by 1.8 to 2.1 MMBD over the entire year.[45]

The industrialized world and the United States help drive this growth in demand. In 2004, the world oil demand increased by 2.7 MMBD, and according to the head of the EIA, a third of the increase is due to the increase in Chinese demand for oil.[46] In 2004, the United States consumed 20.7 MMBD, China consumed 6.5 MMBD, Japan consumed 5.4 MMBD, Germany consumed 2.6 MMBD, Russia consumed 2.3 MMBD, Canada consumed 2.3 MMBD, India consumed 2.3 MMBD, and South Korea consumed 2.1 MMBD.[47]

Oil consumption of the OECD countries in 2004 was 48.8 MMBD, 5.2 percent higher than it was in 2003. The changes in consumptions included the following: United Kingdom (+2.4 percent), Germany (–1.2 percent), Canada (+3.9 percent), France (+0.9 percent), United States (+2.8 percent), Italy (–2.8 percent), Japan (–3.0 percent), and South Korea (–0.8 percent).[48]

World demand for oil in 2002 was 78.20 MMBD, and roughly 83.14 MMBD for the first half of 2005. The EIA estimated that for the first half of 2005, the United States consumed 20.57 MMBD, China consumed 6.81 MMBD, Japan consumed 5.52 MMBD, Germany consumed 2.52 MMBD, Canada consumed 2.35 MMBD, South Korea consumed 2.23 MMBD, and the rest of Asia consumed 8.28 MMBD.[49]

Long-Term Growth by Global Economic Growth Scenario

Longer-term estimates project high growth rates, although partly because total world production capacity of oil is assumed to meet the increases in the global oil demand. Because the *International Energy Outlook 2005* did not examine the impact of high oil prices on world energy balances, there was no way to guess at how much this would change if oil prices remained high. The numbers above, however, provide a good benchmark for the analysis of global oil demand and supply.

In 2005, the EIA provided estimates of the growth in total world demand for three scenarios based on differing rates of economic growth rates: low, reference, and high. For the low economic growth case, world demand is estimated to be 98.60 MMBD in 2010, 110.0 MMBD in 2015, 120.60 MMBD in 2020, and 132.30 MMBD in 2025. For the reference case, total world demand is estimated to be 93.60 MMBD in 2010, 103.20 MMBD in 2015, 111.0 MMBD in 2020, and 119.20 in 2025. For the high economic growth case, total world demand is estimated to reach 91.0 MMBD in 2010, 97.20 MMBD in 2015, 102.30 MMBD in 2020, and 107.7 MMBD in 2025.

According to these projections, the elasticity of demand will become steeper with time. If the economic growth rate changed from reference to high, total world demand for oil would increase by 5.3 percent in 2010 and by 11 percent in 2025; if the economic growth rate went from reference to low, world demand for oil would decrease by 2.8 percent in 2010 and by 9.6 percent in 2025; in the event that the economic growth rate went from low to high, world oil demand would increase by 8.35 percent in 2010 and by 22.8 percent in 2025.[50]

The International Monetary Fund (IMF) projects that world oil demand growth rate will be 2.1 MMBD every year. Due to this surge, the IMF forecasts that the price per barrel of oil will be $34 in 2010 and $39–$56 in 2030. The Fund concludes that the world needs to adapt to high oil prices for the next 20 years and that the global economy faces "permanent oil shock."[51]

Long-Term Growth by Consuming Country

The *International Energy Outlook 2005* projected that high levels of growth in key emerging economies, particularly economies in Asia, such as India and China, will account for much of the increase in global demand for oil, which the EIA projected to grow at 3.5 percent a year over the next 20 years. This growth in the demand for oil is, however, directly linked to robust global economic conditions and particularly in emerging economies such as China and India. The EIA projects the following:

- Transitional economies such as Eastern Europe and the FSU will witness an oil demand growth of 1.4 percent a year, which will translate into an increase in their oil consumption from 5.5 MMBD in 2002 to 7.6 MMBD in 2025.[52]

- China replaced Japan as the second largest consumer of petroleum in 2004. During the same year, total Chinese petroleum products consumption averaged 6.5 MMBD. The EIA claims that China is the source of around 40 percent of the global energy demand growth in 2004.

- The "China factor" will continue to play a major part in global energy demand. Since 1983, China has been a net importer of oil, and it will continue to be dependent on foreign oil, namely, Middle Eastern oil, for the foreseeable future. The EIA projected in 2005 that Chinese consumption could reach 14.2 MMBD in 2025 and that Chinese imports will reach 10.98 MMBD of its total petroleum demand (77 percent of its total consumption needs).[53]

- The "India factor" is also an important factor in shaping the increase in demand. India in 2001 consumed 2.1 MMBD and 2.2 MMBD in 2003. Oil already composes 30 percent of India's energy consumption, but the country has only 5.4 billion barrels of oil.[54] According to the EIA's reference case forecast Indian consumption will reach 2.67 MMBD in 2010 and double to as high as 4.9 MMBD in 2025.[55]

- Growth seems likely to be much slower in developed Asia, but still is an important force in the market. Japan consumed 5.3 MMBD in 1990, 5.4 MMBD in 2001, 5.3 MMBD in 2002, 5.5 MMBD in 2003, and 5.4 MMBD in 2004. The lack of growth in the Japanese demand for oil is also apparent in the EIA forecast. The reference-case forecast of the *International Energy Outlook 2005,* for example, projects that Japan's consumption in 2025 will be 5.3 MMBD.[56]

- Other industrialized nations are projected to have higher growth rates than Japan in spite of rising prices. On April 5, 2005, Alan Greenspan, the Federal Reserve Chairman, said, "Higher prices in recent months have slowed the growth of oil demand, but only modestly." Greenspan also noted that the high oil prices are due to "geopolitical uncertainties" in the oil-producing states. He also argued that "the status of world refining capacity has become worrisome" and that these factors are creating a "price frenzy."[57]

It is important to note that the total projected growth for other Asia nearly totals the growth projected for China. Total Asian demand is estimated to rise from 21.5 MMBD in 2002 to 35.6–45.50 MMBD in 2025. India's oil demand is estimated to increase from 2.20 MMBD in 2002 to 4.30–5.40 MMBD in 2025. Other Asia's oil demand is estimated to rise from 5.6 MMBD in 2002 to 9.80–13.4 MMBD in 2025. South Korea's demand for oil is estimated to rise from 2.20 MMBD in 2002 to 2.60–3.40 MMBD in 2025.

The Forces That Could Limit Oil Supply

If these levels of sustained average demand growth actually occur, virtually all forecasts indicate that it will put a growing strain on both global petroleum supply and export capacity. The BP's *Statistical Review of World Energy 2005* reported that in 2004, the average total world production was 80.26 MMBD—higher than the 2003 average by 3.206 MMBD.

In 2004, OPEC produced 32.927 MMBD, which is a 7.7-percent increase from its 2003 production levels of 2.241 MMBD, Russia increased its production by 0.741 MMBD (+8.9 percent), and China by 0.089 MMBD (+2.9 percent).[58]

Non-OPEC supply has been slow to respond to the high oil prices. In fact, it increased by only 0.046 MMBD in 2004 (31.8 percent of which came from the FSU). According to the U.S. Department of Energy, the increase in non-OPEC oil production for 2005 was 0.92 MMBD.[59] In the years of 2005 and 2006, more than half of this non-OPEC increase was estimated to come from the FSU and the Atlantic Basin, including Latin America and West Africa.[60]

Estimates of spare capacity are increasingly uncertain and inevitably differ. According to the IEA, in early 2005, OPEC had 1.92–2.42 MMBD spare capacity, but according to the EIA, it had 1.1–1.6 MMBD. In both cases, practically all of the spare capacity was from Saudi Arabia.[61]

Longer-Term Supply and Demand

As for longer-term supply and demand, the EIA reported that world production is estimated to increase steadily in the next two decades. Its 2005 modeling efforts estimate marked a slight decrease from the 2004 projections due to the forecasting of a longer period of sustained high oil prices.[62]

In 2002, the world oil production capacity was 80.0 MMBD. Looking toward the future, the EIA estimated world production capacity for the low-price case ($21/barrel), the reference case ($35/barrel), and the high-price case ($48/barrel), and they projected the following:

- For the high-price case, total world oil production capacity is estimated to be 101.60 MMBD in 2010, 113.30 MMBD in 2015, 123.90 MMBD in 2020, and 135.20 MMBD in 2025.

- For the reference case, total world production capacity is estimated to be 96.50 MMBD in 2010, 105.40 MMBD in 2015, 113.60 MMBD in 2020, and 122.20 MMBD in 2025.

- For the low-price case, total world production is estimated to be 94.60 MMBD in 2010, 101.80 MMBD in 2015, 108.50 MMBD in 2020, and 115.50 MMBD in 2025.

As is the case with the elasticity of demand, the elasticity of supply becomes greater with time. Total world oil production capacity will increase by 10.64 percent in 2025 and by 5.28 percent in 2010 if the price changes from reference to high, it will drop by 5.48 percent in 2025 and by 1.97 percent in 2010 if the price changes from reference to low, and it will increase by 17 percent in 2025 and by 7.4 percent in 2010 if the price changes from low to high. In any case, the forecast increases in oil capacity meet the increase in oil demand.[63]

Major Risks and Uncertainties

There are, however, many risks and uncertainties on the supply side in both the short- and the long-term that such forecasts only partially consider. In the short-term, such uncertainties include the following factors:

- The surge in oil demand has pushed many producing countries to produce at their maximum capacity, which instilled fears of the lack of spare capacity in case of further spikes in the market.
- Some oil firms have downgraded their reserve estimates of certain oil fields.
- Oman has falling production levels.
- Kuwait and the United Arab Emirates have been slow to modernize production facilities and techniques.

- The world oil market is losing 1.0 MMBD from depletion every year.[64]
- Uncertainty about the flow of Iraqi oil exports in the face of the high level of internal turmoil and the lack of any infrastructure of technological upgrade in the oil infrastructure since the Gulf War.
- Continued political uncertainties in Iran and unrealistic policies toward foreign investment by the current Iranian government.
- Damage inflicted on U.S. Gulf Coast and offshore oil installations following hurricanes Charley, Frances, Ivan, Katrina, and Rita.
- Capacity constraints (upstream, downstream, and transportation).
- In addition, Venezuelan political instability, Nigerian labor strikes, and internal strife between the Russian government and oil giant, Yukos, also contributed to push crude oil and other petroleum prices higher in 2005.

In the longer-term, such uncertainties include the factors that follow:

- The actual level of producible reserves in virtually all developing states at given levels of price and technology. Experts like Simmons argue that current estimates seriously exaggerate such capability. The country-by-country analyses of the EIA indicate that major additional proven reserves await discovery in Saudi Arabia and virtually every MENA country.
- The real-world cost of incremental production capacity. Current EIA, IEA, and OPEC estimates almost certainly use cost estimates that are too low for Saudi Arabia and other MENA countries and that understate the full cost of infrastructure and advanced recovery techniques. What is not clear is what the real cost will be.
- Debates over the commercially recoverable oil in existing oil fields and countries, the sustainability of production with current recovery techniques, and future technology gain.
- The rate of maturity and decline in given oil fields with present and future technology.
- The future commercial potential of tar sands and heavy oil is a factor that could sharply change the distribution of the world's commercial reserves, if resources like Canadian tar sands become as cost-effective as nations like Canada hope.
- Major uncertainties over the ability to find and produce oil beyond the levels counted in proven reserves. The 2005 BP *Statistical Review* estimates that global proven reserves are 1,188.6 billion barrels,[65] while in 2004 the EIA projected that reserve growth will provide another 334.5 billion barrels (a growth of 26 percent), and there is another 538.4 billion barrels that is undiscovered (an additional growth of 43 percent).[66]
- Long-term substitution effects that bring alternative fuels on-line at competitive prices at whatever petroleum price levels emerge over time.

Supply Disruptions: Real and Perceived

Supply disruptions continue to be a constant risk and have contributed to the high oil prices in recent years. The global energy market has experienced supply disruptions due to labor strikes, oil infrastructure sabotages, and natural disasters. On April 5, 2002, half of the workers in Venezuela's national oil company, Petróleos de

Venezuela SA, went on strike, causing two out of the five Venezuelan export terminals to stop operating.

Another example is provided by the effect of Hurricanes Ivan, Katrina, and Rita. On September 14, 2004, the companies in the Gulf Coast, including Shell, Exxon-Mobil, ChevronTexaco, and TOTAL, shut down their production and evacuated 3,000 of their workers from their offshore platforms. The U.S. Mineral Management Service (MMS) estimated that Ivan caused the Gulf of Mexico's oil production to decline by 61 percent.[67] According to MMS, Hurricane Katrina caused the Gulf Coast production to decline by 0.89 MMBD, which represents roughly 60 percent of the Gulf Coast production.[68]

Capacity constraints, and the perception that supply is limited, have had as much influence in the past as actual supply disruptions. The world energy market will add only 300,000 barrels of net, new, on-stream supply from 2006 to 2010. This lack of growth is met by a 2.5-percent increase in demand. Prices, therefore, would need to rise to clear the market.[69]

The marginal cost of surplus capacity in a high demand market could also become much higher, particularly since the real-world marginal cost of incremental production is rising all over the world because of increased technical sophistication in production and lower-yield oil fields.

All of these factors confront the global economy and energy producers with a world in which demand-driven capacity and export forecasts are not only unreliable, but where the risks caused by such uncertainties are growing more serious. Forecasts based on low prices have already proven wrong, and this may prove equally true of forecasts that point to relatively "high" oil prices that range from $40 to $105.

If such high prices occur, they will eventually dampen demand for crude oil, and the magnitude of the real-world drop depends on the elasticity of demand one assumes. Updated forecast models need to be built to adjust for the recent high oil prices and to modify past assumptions about the interdependence between supply forecasts, prices, and current and future demand.

Demand will always be equally unpredictable. Demand may well decrease with a slowdown in Asian growth: "trees do not grow to the sky" even in China and India. If demand does rise steadily, however, producing countries such as Saudi Arabia will face growing challenges in trying to simultaneously increase production to meet demand, replace depleted fields, and recover a reserve of 2.0 MMBD—which would in any case become a steadily smaller percentage of world demand.

THE GROWING IMPORTANCE OF MENA GAS TO THE GLOBAL MARKET

As noted earlier, while oil demand is projected to increase, the world demand for natural gas is estimated to rise even faster. At present, Middle Eastern gas reserves are more important as a means of meeting local energy needs and reducing domestic MENA consumption of crude oil than as a source of global energy exports. This may change in the future, however, as world demand for gas rises and gas is used more often to provide the raw material for gas-based petrochemicals.

The EIA estimated that global demand for natural gas increased from 36 trillion cubic feet in 1970, to 53 Tcf in 1980, 73.4 Tcf in 1990, 87.0 Tcf in 2000, 89.3 Tcf in 2001, and 92.2 Tcf in 2002. It projects demands to increase to 139.5–173.4 Tcf in 2025, a total some 51 to 88 percent higher than in 2002.[70]

MENA Gas Reserves

There is no firm consensus as to how to estimate proven gas reserves. Many MENA countries are just beginning to make serious efforts to fully characterize their gas reserves and have little experience with the costs of real-world development of massive gas production efforts and the necessary distribution systems. Estimates of proven gas reserves are significantly more controversial than estimates of proven oil reserves, and estimates of potential and undiscovered reserves are too uncertain to be used for the purposes of this analysis.

BP Estimates of Trends in Gas Reserves

Figure 1.22 shows BP estimates of regional proven gas reserves. In 2004, the Middle East and North Africa had a total of 44.9 percent of the world's proven gas reserves (40.6 percent in the Middle East and the rest in Algeria, Egypt, and Libya). BP estimates these MENA reserves increased from 967.28 Tcf in 1984, to 1,608 Tcf in 1994, and to 2,570 Tcf in 2004. This is an increase of 165 percent over two decades.[71]

Figure 1.23 shows the recent trend in estimates of MENA gas reserves by country and the growth in estimates of the resources of many countries. It provides a graphical representation of how estimates of the MENA region's proven gas reserves have steadily grown in the past two decades.

The BP estimates provide the following country-by-country breakdown of the trends in Middle Eastern proven natural gas reserves:[72]

- Bahrain is the only country in the MENA region that has witnessed a decrease in its proven gas reserves from 7.5 Tcf to 3.17 Tcf, a 57-percent decline from 1984 to 2004.
- Iran is estimated to have increased its proven gas reserves from 494.7 Tcf in 1984 to 970 Tcf in 2004, a 96.2-percent increase.
- Iraq is estimated to have increased its proven gas reserves from 28.8 Tcf in 1984 to 111.9 Tcf in 2004, a 287-percent increase.
- Kuwait is estimated to have increased its gas proven gas reserves from 36.6 Tcf in 1984 to 55.49 Tcf in 2004, a 51.5-percent increase.
- Oman is estimated to have increased its proven gas reserves from 7.58 Tcf in 1984 to 35.1 Tcf in 2004, a 362-percent increase.
- Qatar is estimated to have increased its proven gas reserves from 151.08 Tcf in 1984 to 910.13 Tcf in 2004, a 502-percent increase.

Figure 1.22   BP Estimates of MENA and Other Regional Proven Gas Reserves: 1984–2004 (in Tcf)

Nation	1984	1994	2004	% Change 1984–2004	% of World Total
Bahrain	7.552	5.40	3.17	−57.9%	0.10%
Iran	494.76	732.96	970.75	96.2%	15.30%
Iraq	28.87	109.95	111.90	287.5%	1.80%
Kuwait	36.63	52.92	55.49	51.5%	0.90%
Oman	7.58	9.03	35.12	362.8%	0.60%
Qatar	151.08	249.57	910.13	502.4%	14.40%
Saudi Arabia	127.36	185.67	238.41	87.2%	3.80%
Syria	3.67	8.29	13.09	256.7%	0.20%
UAE	109.71	239.22	213.9	95.0%	3.40%
Yemen	–	15.14	16.90	?	0.30%
Other	+	+	1.87	?	<0.05%
Total Middle East	**967.28**	**1,608.419**	**2,570.79**	**165.8%**	**40.60%**
Algeria	121.50	104.5939	160.43	32.0%	2.50%
Egypt	8.29	22.2743	65.45	689.0%	1.00%
Libya	22.16	46.243	52.63	137.4%	0.80%
Total MENA	**1,119.25**	**1,781.53**	**2,849.316**	**155%**	**44.90%**
Other Africa	67.423	149.2131	217.90	223.2%	3.50%
Asia Pacific	247.91	355.3051	501.51	102.3%	7.90%
Europe/Eurasia	1,483.14	2,254.752	2,259.73	52.4%	35.70%
North America	371.00	297.312	258.29	−30.4%	4.10%
S. & C. America	113.99	205.7302	250.59	119.8%	4.00%
World Total	**3,402**	**5,043**	**6,337**	**86.2%**	**100.00%**

Note: + means that the proven reserves are estimated to be less than 0.05 Tcf.
Source: BP, *Statistical Review of World Energy 2005.*

- Saudi Arabia is estimated to have increased its proven gas reserves from 127.36 Tcf in 1984 to 238.41 Tcf in 2004, an 87-percent increase.

- Syria is estimated to have increased its proven gas reserves from 3.6 Tcf in 1984 to 13.09 Tcf in 2004, a 256-percent increase.

- United Arab Emirates is estimated to have increased its proven gas reserves from 109.71 Tcf in 1984 to 213.9 Tcf in 2004, a 95-percent increase.

The BP estimates provide a country-by-country breakdown of the North African natural gas reserves and estimated that the North African proven gas reserves grew by 83 percent:[73]

- The MENA country that has seen the largest growth in its proven gas reserves is Egypt. In 1984, it was estimated to have has 8.29 Tcf. The estimate increased to 65.45 Tcf in 2004, a 689-percent increase in 20 years.

- Algeria is estimated to have increased its proven gas reserves from 121.5 Tcf in 1984 to 160.4 Tcf in 2004, an 87-percent increase.

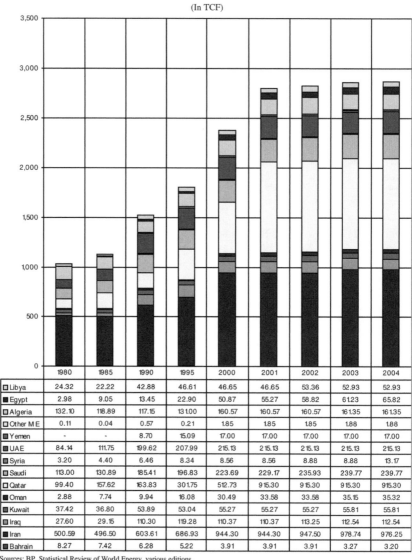

(In TCF)

	1980	1985	1990	1995	2000	2001	2002	2003	2004
☐ Libya	24.32	22.22	42.88	46.61	46.65	46.65	53.36	52.93	52.93
■ Egypt	2.98	9.05	13.45	22.90	50.87	55.27	58.82	61.23	65.82
☐ Algeria	132.10	118.89	117.15	131.00	160.57	160.57	160.57	161.35	161.35
☐ Other ME	0.11	0.04	0.57	0.21	1.85	1.85	1.85	1.88	1.88
■ Yemen	-	-	8.70	15.09	17.00	17.00	17.00	17.00	17.00
■ UAE	84.14	111.75	199.62	207.99	215.13	215.13	215.13	215.13	215.13
■ Syria	3.20	4.40	6.46	8.34	8.56	8.56	8.88	8.88	13.17
☐ Saudi	113.00	130.89	185.41	196.83	223.69	229.17	235.93	239.77	239.77
☐ Qatar	99.40	157.62	163.83	301.75	512.73	915.30	915.30	915.30	915.30
■ Oman	2.88	7.74	9.94	16.08	30.49	33.58	33.58	35.15	35.32
■ Kuwait	37.42	36.80	53.89	53.04	55.27	55.27	55.27	55.81	55.81
☐ Iraq	27.60	29.15	110.30	119.28	110.37	110.37	113.25	112.54	112.54
■ Iran	500.59	496.50	603.61	686.93	944.30	944.30	947.50	978.74	976.25
■ Bahrain	8.27	7.42	6.28	5.22	3.91	3.91	3.91	3.27	3.20

Sources: BP, Statistical Review of World Energy, various editions.

Figure 1.23   BP Estimates of Total Proven Gas Reserves of the MENA States 1979–2004

- Libya is estimated to have increased its proven gas reserves from 22.16 Tcf in 1984 to 52.6 Tcf in 2004, a 137.4-percent increase.

EIA and IEA Estimates of Trends in Gas Reserves

The EIA and the IEA do not provide detailed projections of probable discoveries of new gas reserves by country, and they use somewhat different definitions of gas reserves than BP. They both do, however, report that many Middle Eastern states have only begun to fully explore their gas reserves and that most are likely to make major additional discoveries.

The EIA reports that total global reserves now total 5,501 Tcf, and undiscovered reserves total another 4,839 Tcf—almost all in the developing world. Some 2,347 Tcf in reserves are expected to be discovered during 2000–2025, and more than one-half is estimated to be found in the FSU and MENA areas. If these estimates are right, the Middle East and North Africa have 20 to 25 percent of the world's undiscovered reserves.[74]

The IEA estimates that the Middle East has more than 1,383 Tcf undiscovered gas reserves, 2,869 Tcf of remaining reserves, and 4,353 Tcf of remaining ultimately recoverable reserves. In addition, it estimates that the MENA region has 45 percent of the world reserves and 14.2 percent of the world production and a reserves-to-production ratio of 211 compared to that of the world at large of 66. The IEA estimates the proven natural reserves for the MENA countries—relative to the world's 6,354 Tcf—as follows:[75]

- Iran has 16 percent of the world's gas proven reserves with nearly 995 Tcf.
- Iraq has 2 percent of the world's gas proven reserves with nearly 109 Tcf.
- Kuwait has 1 percent of the world's gas proven reserves with nearly 56 Tcf.
- Qatar has 14 percent of the world's gas proven reserves with nearly 910 Tcf.
- Saudi Arabia has 4 percent of the world's gas proven reserves with nearly 236 Tcf.
- United Arab Emirates has 3 percent of the world's gas proven reserves with nearly 215 Tcf.
- Algeria has 3 percent of the world's gas proven reserves with nearly 162 Tcf.
- Egypt has 1 percent of the world's gas proven reserves with nearly 67 Tcf.
- Libya has 1 percent of the world's gas proven reserves with nearly 53 Tcf.
- Other MENA countries have 1 percent of the world's gas proven reserves with nearly 70 Tcf.

In contrast, most sources agree that the United States, one of the world's largest gas consumers, has now less than 10 percent of the world's remaining reserves. This means the United States will become steadily more dependent on imports—largely from Canada and Mexico. Europe is one of the fastest growing consumers of gas, but is depleting its reserves and will become steadily more dependent on imports from the FSU and the MENA region. Some sources indicate that Europe will have to import 60 percent of its natural gas by 2020.[76] Japan and most developing Asian states have little or no significant reserves.

MENA Gas Consumption

One key aspect of MENA gas reserves is that they give many MENA exporting countries the ability to limit the growth of domestic consumption of crude oil.

According to the EIA, between 1990 and 2002, the Middle East consumption of oil increased from 3.8 MMBD to 5.7 MMBD, a 50 percent increase.[77]

This consumption of oil would have been far greater if Middle East oil exporters had not steadily increased their use of local gas as a substitute for oil. During the same period, however, the Middle East consumption of natural gas increased from 3.7 Tcf to 8.3 Tcf, a 124 percent increase. BP estimates that during the same period, the MENA total natural gas consumption increased by 107 percent, from 12.1 billion cubic feet (BCF) to 25.1 BCF.[78] This increase was driven by the creation of more effective national gas distribution systems in key exporters like Iran, Kuwait, and Saudi Arabia.[79]

The EIA *International Energy Outlook 2005* did not provide a country breakdown of where this increase came from, but the BP provides a country-by-country consumption patterns analysis. From 1994 to 2004, BP *Statistical Review of World Energy 2005* estimated that the MENA total natural gas consumption increased by 80 percent, from 15.5 BCF to 27.9 BCF.

The following national increases took place in MENA natural gas consumption during this period:[80]

- Iran's natural gas consumption increased by 171 percent, from 3.1 BCF in 1994 to 8.4 BCF in 2004.

- Kuwait's natural gas consumption increased by 50 percent, from 0.6 BCF in 1994 to 0.9 BCF in 2004.

- Qatar's natural gas consumption increased by 15 percent, from 1.3 BCF in 1994 to 1.5 BCF in 2004.

- Saudi's natural gas consumption increased by 51 percent, from 4.1 BCF in 1994 to 6.2 BCF in 2004.

- United Arab Emirates' natural gas consumption increased by 81 percent, from 2.1 BCF in 1994 to 3.8 BCF in 2004.

- Algeria's natural gas consumption increased by 5 percent, from 1.9 BCF in 1994 to 2.0 BCF in 2004.

- Egypt's natural gas consumption increased by 150 percent, from 1.0 BCF in 1994 to 2.5 BCF in 2004.

The IEA projected that total Middle Eastern use of natural gas will increase from 3.6 Tcf in 1990, and 6.8 Tcf in 2000, to 8.8 Tcf in 2010, 11.1 Tcf in 2020, and 13.9 Tcf in 2025. This is an average annual increase in consumption of 2.3 percent.[81]

The EIA 2005 reference-case estimates, on the other hand, projected that Middle Eastern natural gas consumption will rise from 7.6 Tcf in 2001 and 8.3 Tcf in 2002, to 10.6 Tcf in 2010, to 12.6 Tcf in 2015, to 14.5 Tcf in 2020, and to 16.6 Tcf in 2025. From 2002 to 2025, the Middle Eastern domestic consumption of gas will increase at an average rate of 3.1 percent per year.[82]

In 2005, country plans called for major further increases in domestic use of gas in most Gulf States. Like the low costs of oil production in these states, this increased domestic use of gas is a major reason they are expected to be able to make major increases in oil exports.

MENA Gas Production

Increases in MENA gas exports are also projected to be nearly as important to the global gas market as the increases in MENA oil exports. MENA gas production has not played a critical role in the world energy market up until now. In 2004, BP estimated that the MENA region produced only 10.48 percent of the world total production.

As Figure 1.24 shows, however, MENA production grew tremendously between 1994 and 2004. Its growth is second to that of Africa in percentage terms, but much more significant in terms of production levels. MENA gas production grew by 306 percent between 1994 and 2004. Africa's grew by 600 percent. However, the MENA region produced 27 BCF compared to 2.8 BCF for Africa. The only region with higher production than the MENA region is Europe/Eurasia. It has a total production of 101.5 BCF, dominated by Russian gas production. Total European/

Figure 1.24   BP Estimates of Regional Gas Production Trends: 1984–2004 (in BCF)

Nation	1984	1994	2004	2004	% of World Total
Bahrain	0.4	0.7	0.9	125%	0.35%
Iran	1.3	3.1	8.2	531%	3.17%
Kuwait	0.4	0.6	0.9	125%	0.35%
Oman	0.1	0.3	1.7	1600%	0.66%
Qatar	0.6	1.3	3.8	533%	1.47%
Saudi Arabia	1.8	4.1	6.2	244%	2.40%
Syria	–	0.1	0.5	–	0.19%
UAE	1.1	2.5	4.4	300%	1.70%
Other	0.1	0.3	0.3	200%	0.12%
Total Middle East	**5.7**	**13.0**	**27.0**	**374%**	**10.43%**
Algeria	3.0	5.0	7.9	163%	3.05%
Egypt	0.3	1.0	2.6	767%	1.00%
Libya	0.4	0.6	0.7	75%	0.27%
Total MENA	**9.4**	**19.6**	**38.2**	**306%**	**14.76%**
Other Africa	0.4	0.7	2.8	600%	1.08%
Asia Pacific	9.3	19.3	31.2	235%	12.06%
Europe/Eurasia	74.3	87.8	101.5	37%	39.22%
North America	58.1	68.5	72.6	25%	28.05%
S. & C. America	4.3	6.5	12.5	191%	4.83%
World Total	**155.9**	**202.5**	**258.8**	**66%**	**100.00%**

Source: BP, *Statistical Review of World Energy 2005.*

Eurasian reserves, however, total 2,259.7 Tcf vs. 2.849.3 Tcf for the MENA region.

To understand the historical and future trends in production, it is necessary to examine what has happened on a country-by-country basis. The following figures provide an analytic assessment of the BP estimates of the MENA natural gas production capacity:[83]

- Figure 1.24 compares the growth of other regions to that of the MENA region from 1984 to 2004. It shows the steady increase in MENA gas production in recent years and how the MENA region dominated the growth. The Middle Eastern countries grew by 374 percent, while total MENA grew by 306 percent between 1984 and 2004. The total world production grew by 66 percent, from 155.6 BCF to 258.8 BCF.

- Figure 1.25 provides a graphical representation of how the MENA gas production has increased since 1970. The Gulf gas production, for example, grew from a mere 2.0 BCF in 1970 to 20.0 BCF in 2004—a tenfold increase during this period.

Looking toward the future, the IEA *World Energy Outlook 2005* estimated the MENA gas production capacity to increase from 13,590 BCF in 2003 to 42,748 Tcf in 2030, a 214-percent increase in MENA's total gas production capacity. Figure 1.26 shows the IEA estimate of country-by-country increases during the projected period. It also highlights the importance of Iran, Qatar, Algeria, and Saudi Arabia within the MENA countries and as the leading gas producers in the world.

The IEA 2005 estimates called for a large increase in Gulf and MENA production. Gulf gas production is estimated to increase by approximately 242 percent during 2003–2030, from 8,103 Tcf in 2003 to 27,463 Tcf in 2030. As is the case with IEA projections of oil production, however, it is uncertain if these estimates will be met. Countries in the region are investing substantial resources into gas exploration, research and development, exploration and production technologies, and best practices. However, the impact such investments will ultimately have on overall gas production capacity is uncertain.

The EIA summarized its 2005 estimates of the MENA natural gas production and trades as follows:[84]

MENA gas production is projected to treble over the projection period from 385 bcm [billion cubic meters] in 2003 to 1211 bcm in 2030. The rate of growth will be faster than that of any other major world region. The biggest volume increases in the region will occur in Qatar, Iran, Algeria and Saudi Arabia.

The call on MENA gas supply will increase rapidly over the projection period, a result of strong global demand and declining output in many other gas-producing regions. Net exports from MENA countries to other regions are projected to climb from 97 bcm in 2003 to 444 bcm in 2030. Most of the increase will be in the form of LNG [liquid natural gas]. There will be a marked shift in the balance of Middle East exports from eastern to western markets. Europe will remain the primary destination for North African gas exports.

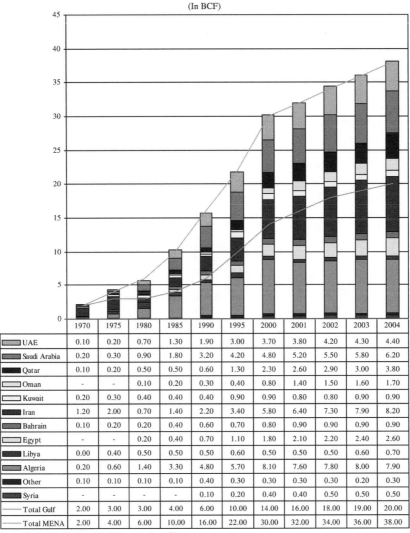

(In BCF)

	1970	1975	1980	1985	1990	1995	2000	2001	2002	2003	2004
UAE	0.10	0.20	0.70	1.30	1.90	3.00	3.70	3.80	4.20	4.30	4.40
Saudi Arabia	0.20	0.30	0.90	1.80	3.20	4.20	4.80	5.20	5.50	5.80	6.20
Qatar	0.10	0.20	0.50	0.50	0.60	1.30	2.30	2.60	2.90	3.00	3.80
Oman	-	-	0.10	0.20	0.30	0.40	0.80	1.40	1.50	1.60	1.70
Kuwait	0.20	0.30	0.40	0.40	0.40	0.90	0.90	0.80	0.80	0.90	0.90
Iran	1.20	2.00	0.70	1.40	2.20	3.40	5.80	6.40	7.30	7.90	8.20
Bahrain	0.10	0.20	0.20	0.40	0.60	0.70	0.80	0.90	0.90	0.90	0.90
Egypt	-	-	0.20	0.40	0.70	1.10	1.80	2.10	2.20	2.40	2.60
Libya	0.00	0.40	0.50	0.50	0.50	0.60	0.50	0.50	0.50	0.60	0.70
Algeria	0.20	0.60	1.40	3.30	4.80	5.70	8.10	7.60	7.80	8.00	7.90
Other	0.10	0.10	0.10	0.10	0.40	0.30	0.30	0.30	0.30	0.20	0.30
Syria	-	-	-	-	0.10	0.20	0.40	0.40	0.50	0.50	0.50
Total Gulf	2.00	3.00	3.00	4.00	6.00	10.00	14.00	16.00	18.00	19.00	20.00
Total MENA	2.00	4.00	6.00	10.00	16.00	22.00	30.00	32.00	34.00	36.00	38.00

Source: BP, Statistical Review of World Energy, various editions.

Figure 1.25 BP Estimates of MENA Natural Gas Production by Country: 1970–2004

The surge in energy demand across fuel types, the high oil prices, and the call by consumer countries for more diversification in their consumption portfolios, have increased the importance of natural gas in the energy market. This is true for both the MENA domestic consumption as well as the world's energy demand.

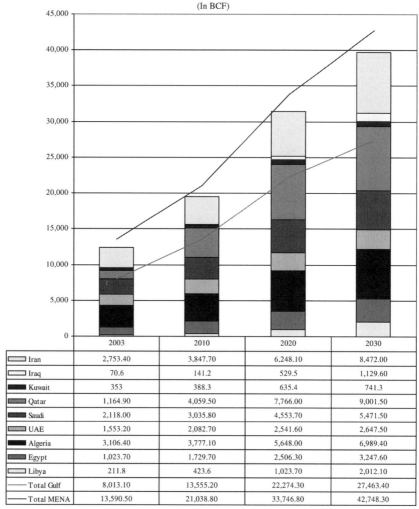

	2003	2010	2020	2030
Iran	2,753.40	3,847.70	6,248.10	8,472.00
Iraq	70.6	141.2	529.5	1,129.60
Kuwait	353	388.3	635.4	741.3
Qatar	1,164.90	4,059.50	7,766.00	9,001.50
Saudi	2,118.00	3,035.80	4,553.70	5,471.50
UAE	1,553.20	2,082.70	2,541.60	2,647.50
Algeria	3,106.40	3,777.10	5,648.00	6,989.40
Egypt	1,023.70	1,729.70	2,506.30	3,247.60
Libya	211.8	423.6	1,023.70	2,012.10
Total Gulf	8,013.10	13,555.20	22,274.30	27,463.40
Total MENA	13,590.50	21,038.80	33,746.80	42,748.30

Source: IEA, World Energy Outlook 2005, Middle East and North Africa Insights.

Figure 1.26   IEA Estimates of MENA Natural Gas Production by Country: 2003–2030

MENA Gas Exports

An analysis of the MENA region's future role in gas exports is necessarily more speculative than an analysis of its role in oil exports. While the MENA area has long exported some gas, these exports are just beginning to become a major part of world energy exports, and projections must be based on highly uncertain data as to future export capacity, future demand, and future price.

- The EIA and IEA do, however, project world demand for gas as one of the most rapidly growing areas of energy demand. The reference case of the EIA 2005 estimates projected that world use of gas will rise from 7.4 Tcf in 1990, 89.3 Tcf in 2001, and 92.2 Tcf in 2002, to 111.4 Tcf in 2010, 127.9 Tcf in 2015, 141.6 Tcf in 2020, and 156.6 Tcf in 2025. This is an average annual increase in consumption of 2.3 percent vs. 1.9 percent for oil.[85]

- The IEA's reference scenario for 2005 projected a similar level of increases. It estimated that the world's demand for natural gas would increase from 95.6 Tcf in 2003, to 113.4 Tcf in 2010, to 143.3 Tcf in 2020, to 169.0 Tcf in 2030. This is an average increase of 2.1 percent.[86]

- The countries with the largest projected increases in gas demand are China, Brazil, and India. According to the EIA reference-case projections, China's demand for natural gas is estimated to increase by an average of 7.8 percent a year between 2002 and 2025. The EIA estimates China's demand for natural gas is to increase from 1.2 Tcf in 2002 to 2.6 Tcf in 2010, to 3.4 Tcf in 2015, to 4.2 Tcf in 2020, and to 33.3 Tcf in 2025.

- India is estimated to increase its demand for natural gas from 0.9 Tcf in 2020, to 1.4 Tcf in 2010, to 3.4 Tcf in 2015, to 4.2 Tcf in 2020, to 6.5 Tcf in 2025—a 5.1-percent average annual change from 2002 to 2025. Brazil is also estimated to increase its gas consumption by 6.8 percent a year between 2002 and 2025. The EIA estimates that Brazil's demand will increase from 0.5 Tcf in 2002 to 0.9 Tcf in 2010, to 1.3 Tcf in 2015, to 1.7 Tcf in 2020, to 2.1 Tcf in 2025.[87]

- Key existing gas users like the United States will almost certainly have to make major increases in gas imports by tanker. In addition, Korea and Japan already rely heavily on tankers to deliver MENA gas exports. In the EIA reference-case projection, the U.S. natural gas consumption is estimated to increase at an average rate of 1.5 percent between 2002 and 2025, or from 19.2 Tcf in 1990 and 23.0 Tcf in 2002, to 25.6 Tcf in 2010, to 28.3 Tcf in 2015, to 30.4 Tcf in 2020, to 30.9 Tcf in 2025.

- Western European demand for natural gas is estimated to increase from 10.1 Tcf in 1990, to 14.9 Tcf in 2001, to 15.0 Tcf in 2002, to 17.3 Tcf in 2010, to 19.0 Tcf in 2015, to 20.4 Tcf in 2020, to 22.4 Tcf in 2025—an average annual increase of 1.8 percent between 2002 and 2025.[88]

The IEA estimates the future growth in MENA natural gas exports between 2003 and 2030. Figure 1.27 shows the IEA estimates of the upward trends in MENA gas net exports between 2003 and 2030. It estimates that total MENA net exports will increase from 3,424 Tcf in 2003, to 6,636 Tcf in 2010, to 11,543 Tcf in 2020, to 15,673 Tcf in 2030—a 357-percent increase from 2003 to 2030. The Gulf States are estimated to increase their production by 757 percent during the same period, from 988.4 Tcf in 2003 to 8,472 Tcf in 2030.[89]

For these increases to take place, the MENA countries, Russia, and the other FSU states will have to sharply increase their gas production and export capacities. At present, however, MENA gas production lags far behind the level of production by

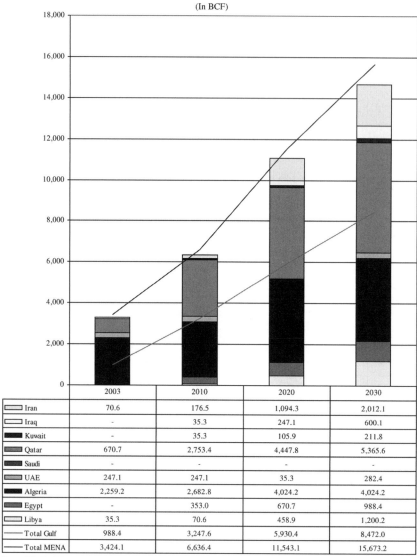

	2003	2010	2020	2030
Iran	70.6	176.5	1,094.3	2,012.1
Iraq	-	35.3	247.1	600.1
Kuwait	-	35.3	105.9	211.8
Qatar	670.7	2,753.4	4,447.8	5,365.6
Saudi	-	-	-	-
UAE	247.1	247.1	35.3	282.4
Algeria	2,259.2	2,682.8	4,024.2	4,024.2
Egypt	-	353.0	670.7	988.4
Libya	35.3	70.6	458.9	1,200.2
Total Gulf	988.4	3,247.6	5,930.4	8,472.0
Total MENA	3,424.1	6,636.4	11,543.1	15,673.2

Source: IEA, World Energy Outlook 2005, Middle East and North Africa Insights.

Figure 1.27   IEA Estimates of MENA Natural Gas Net Trade by Country: 2003–2030

the FSU and Eastern Europe. The MENA region has one-third of their total production, although the MENA region has slightly larger total gas reserves.

All Gulf gas exports are now in the form of LNG. Iran, however, is exploring shipping gas to Europe by pipeline through Turkey, and several Gulf States have considered pipelines through the Indian Ocean, Pakistan, or to India across Afghanistan.

Qatar is a particularly important potential source of pipeline exports. It has much larger gas reserves than oil reserves and has aggressively expanded its LNG facilities. It is seeking to triple its LNG capacity to 45 million metric tons per year by 2010, plus building new gas-to-liquid plants. It is also a key force behind the creation of the first long-distance pipeline to serve customers in the Gulf area—the Dolphin Project. (The United Arab Emirates' production of gas is largely associated gas and is limited by oil production, and its consumption of gas is outstripping supply.)[90] Saudi Arabia has planned a massive new Gas Initiative, and while its efforts to find foreign investment have been delayed and scaled back—it too is likely to become a significant exporter over the coming years.

The situation is different outside the Gulf. Algeria is already the second largest LNG producer in the world and has significant exports by pipeline. It is western Europe's second-largest supplier of exports and delivers supplies by pipeline to Italy, Spain, and Portugal, and by LNG tanker to France, Spain, Italy, Belgium, Greece, and Portugal. Algeria is seeking to add a new 4-million metric ton LNG train to its production and is trying to diversify exports to new markets in the United States. It exports about 0.8 Tcf via the Transmed pipeline through Tunisia to Italy, and Algeria and Italy are exploring the possibility of a new pipeline through Sardinia and Corsica. Another "Medgaz" pipeline may be built to Spain, with a capacity growing from 0.3 Tcf to 0.6 Tcf.

Egypt is creating gas trains to export to France and Spain, and Libya is planning to increase its gas export capability by building a pipeline from Melita to Sicily with a capacity of 0.3 Tcf.[91] In addition, Egypt has announced plans for gas exports (either by pipeline or liquefied natural gas tanker) to such countries as Turkey, Israel, Jordan, and the Palestinian territories. Pricing issues have complicated these plans—specifically, how much exported gas should cost relative to domestically consumed gas. The International Energy Agency, however, estimated that Egypt may complete five major LNG projects by 2007 with a total cost of $5.4 billion. Two have been completed as of September 2005.[92]

Furthermore, several natural gas production projects are currently under way in Libya, including at as-Sarah and Nahoora, Faregh, Wafa, offshore block NC-41, Abu-Attifel, Intisar, and block NC-98. The Western Libyan Gas Project (WLGP) and the $6.6-billion "Greenstream" underwater gas pipeline came on-line in 2004, resulting in a rapid increase in gas exports to Europe. Although Spain's Engas was originally Libya's only customer, the WLGP (Eni and National Oil Corporation) has facilitated expansion to Italy and eventually several more European states.

Another proposal in North Africa is to build a 900-mile gas pipeline linking Egypt and Libya to Tunisia and Algeria, from where it would hook up with the existing pipeline to Morocco and Spain. Tunisia and Libya already signed a separate agreement for 70 BCF per year of natural gas to be delivered from Libyan gas fields to Cap Bon, Tunisia. In October 2003, the two nations opened a joint venture gas company that will build the pipeline.

DEALING WITH AN UNCERTAIN FUTURE IN THE OIL AND GAS MARKETS

The details of energy often seem boring, particularly compared to exciting speculative debates over energy politics and alternative energy supplies. So do comparisons of statistics, and the results of complex models. However, the moment one actually looks in detail at the numbers projected by the most respected sources of energy data and future estimates, it becomes clear that such details really count. Such estimates and models make it clear that there are no near- or midterm developments in the real world that are likely to eliminate a growing global dependence on Middle Eastern energy exports, or the world's dependence on the ability and willingness of the Middle East to increase its energy production and export capability.

Time and technology will almost certainly change the world's dependence on MENA oil and gas exports, but this will not happen in a few years. Radical changes may take more than a few decades—barring some massive, unanticipated breakthrough in alternative energy supplies. For example, even if dramatic changes do take place in the technology and the production cost of alternative energy supplies, it might well take a decade for such changes to reach the scale necessary to have a decisive global impact. The world has simply invested too much in vehicles, facilities, homes, and industrial processes that use oil, and few breakthroughs could take the form of supplies that could be cheaply and quickly produced on a global basis.

These realities may not be apparent when the numbers that lie behind global energy balances are ignored, or when policy makers and analysts look at only part of the problem—such as the size of today's direct U.S. imports of crude oil from the Middle East. It is clear from the previous analysis, however, that the real world is far more complicated and that any honest analysis must reflect that complexity. It should be equally clear that major changes in the future projected by groups like the IEA and the EIA might change the numbers, but are unlikely to change the broad trend in ways that radically affect the pattern of world energy consumption, world energy exports, or global dependence on the Middle East.

Key Areas of Uncertainty in Estimates of Increases in Production Capacity

It should be stressed that it is unlikely that all the MENA countries will make all of the production increases at the level that the EIA and the IEA estimate and that their estimates of future energy costs may be more representative of what the market wants than what these countries are capable of supplying or willing to sell at such prices.

As has been noted from the start, these projected increases in production capacity are based on economic models that assume MENA states can and will expand production capacity to meet market demand. They are not based on country plans to actually fund and implement such increases. This makes any such estimates—and

THE IMPORTANCE OF MENA ENERGY   63

the related projections of increases in exports—much more uncertain than they would be.

There are dangers in estimating and modeling the behavior of countries before the countries involved have made any clear decisions about their future plans. At the same time, most countries in the region have never made serious long-range plans to expand production capacity, and most have reacted to market forces although few have risked anticipating them.

There are many technical uncertainties in estimating the size and character of oil and gas reserves, and the cost of maintaining and expanding production. Errors of 20 percent or more can easily occur in projecting the mid- and the long-term nature and behavior of a given field, particularly when countries have not updated their exploration, testing, and management techniques. Abu Dhabi, Iran, Iraq, Kuwait, Libya, Oman, and Yemen are all examples of MENA states that have demonstrated in the past that they have questionable capability to accurately characterize their reserves and execute oil field development and increase production on a "best practices" basis, although most of these countries do at least approach international standards.

It is equally important to note that market forces are only one of the factors that have shaped MENA behavior in expanding oil and gas production capacity and reserves. The MENA region has been the scene of more than 10 conflicts and major internal security struggles over the last two decades, and the production capacity and exports of several states have been affected by UN and U.S. sanctions. As a result, it is important to review the assessments of estimated increases in production by country and understand that the expansion in each country involves both economic and security risks.

Simply modeling uncertainty does not mean that the resulting conclusions establish the boundaries of the problem. Many in the oil industry feel that all of the EIA and IEA estimates of future production capacity are too high and that the countries in the region will be much slower to increase production.

It should also be noted that several real-world trends do not conform to the EIA and IEA projections even in the short-term. Iraqi oil production was only 0.8– 1.2 MMBD in August 2003 because of the impact of the Iraq War and its aftermath. Iran and Libya have failed to modernize and increase production for more than half a decade because of internal political developments and external sanctions. Kuwait has fallen badly behind in field development and technology because its National Assembly has blocked suitable investment reforms. Algeria continues a civil war, and the problem of terrorism has become more serious in the Gulf region in general and Saudi Arabia in particular.

This does not necessarily mean that the EIA and IEA projections will not prove broadly accurate over time, but it does mean that there are security as well as market risks and that future production and export capacity is as much an energy risk as embargos or temporary interruptions in production.

At the same time, the sources of current and increased oil production in other regions—the FSU, Latin America, and West Africa—are all subject to the same general uncertainties as those in the MENA area. The countries involved have been at least as affected by poor state planning and development, and conflict and internal instability, as the MENA area. Energy comes from a risk-filled world, and only truly massive and lasting shifts in the pattern of regional oil exports have true strategic importance.

Geopolitical, Strategic, and Security Dimensions in the MENA Region

The many uncertainties affecting the international energy market, the discovery and exploitation of energy reserves, and the competition between fuels are only part of the forces shaping the Middle East energy supply. The analysis in Chapter 1 discussed estimates that assume that market forces will dominate the future development of energy exports in the Middle East and North Africa. The MENA region, however, has been the scene of many internal crises and external conflicts. There have been several past occasions on which these crises have affected either the flow of MENA energy exports or the development of Middle Eastern energy production and export capacity.

The MENA region is anything but stable today, and there is a wide range of external and internal forces that may become major future threats to Middle Eastern energy exports. The politics, economics, and social dynamics that shape these threats are complex. They are driven by political and security issues, but they are also driven by economic and demographic factors and a wide range of cultural factors. It is also dangerous to generalize. The MENA area includes at least 22 nations, located in an arc that sweeps from North Africa to the edge of Central Asia and the Red Sea. As of 2005, these states had a total population of some 348.24 million and a gross domestic product (GDP) of some $561.78 billion, and each had different political, economic, demographic, and security conditions and needs.

Most MENA states are Arab and Muslim, but a common ethnic and religious background has never meant that they do not go to war with each other or do not have internal sectarian, ethnic, and political conflicts. The MENA region is also divided into at least four subregions, each of whose nations have different interests and present different risks. These four subregions include the Maghreb, with Mauritania, Morocco, Algeria, Libya, and Tunisia; the Levant and the Arab-Israeli confrontation states: Egypt, Israel, Jordan, Lebanon, and Syria; the Gulf: Iran, Iraq, Kuwait, Bahrain, Qatar, Saudi Arabia, the United Arab Emirates, and Oman; and the Red Sea states like Yemen, the Sudan, and Somalia.

Each subregion includes states that have been the source of recent conflicts, although many states have been comparatively stable and have regimes with a long history of friendship to the West. The economics of every MENA state relies on strong trading partners outside of the region, and these links also differ by subregion and nation. The nations of North Africa are linked closely to southern Europe and also have ties to the sub-Saharan states. The states of the Levant trade primarily with Europe and the United States. The southern Gulf States trade with the West and increasingly with Asia and the developing world. Iran is in many ways a Central Asian state that exports through the Gulf. It has good reason to be deeply concerned about security issues in Afghanistan and proliferation in India and Pakistan.

Treating the Middle East as a "region," rather than as a group of distinct actors, often conceals far more than it reveals. The future development of energy supply in each nation and subregion will be affected by exporting different political, security, ethnic, and sectarian fault lines. The internal character and strategic interests of given nations differ sharply from state to state, and, as is analyzed in Chapter 3, MENA states differ politically, economically, and demographically in many intricate and important ways. In many cases, regional or national tensions have already led to war and could lead to future conflicts. In other cases, internal tensions have produced civil conflicts. Violent religious extremism is an ongoing problem in many MENA countries, and the events of September 11, 2001, have only dramatized internal security problems and terrorism that has long existed on a national and a regional level.

A HISTORY OF CONFLICT AND TENSION

MENA nations have a long history of violence and conflict. The Arab-Israeli Wars of 1948, 1956, 1967, 1970, 1973, 1982, and the first and the second *Intifadha* are all cases in point. So are the Iran-Iraq War, the Iraqi invasion of Kuwait, the Gulf War, and the Iraq War. There is little chance that the region will avoid new conflicts between the present and 2020. Many Middle Eastern states still dispute at least one border with one of its neighbors, and most countries have serious religious and/or ethnic divisions. Low-level conflicts and internal unrest are virtual certainties.

In several cases, Middle Eastern states are either currently at war, or there is a serious risk of future conflict:

- Mauritania has long been the scene of a low-level race war between Arabs and Black Africans and between the military and the ruling elites.
- Morocco is still in the process of a long war with the Polisario for control of the Western Sahara.
- Algeria is involved in a bitter civil war between its ruling military junta and Islamic extremists.

- Tensions have grown between Libya's leader, Muammar Qadhafi, and Libya's Islamists, and there has been low-level fighting in a number of areas.

- The Egyptian government and a number of other regional governments are fighting low-level wars against extremist groups.

- A "war process," or "Second Intifadha," has replaced the Arab-Israeli peace process. Israel is still formally at war with Syria and Lebanon and faces a serious potential threat from outside terrorists. Israel may become involved in a broader conflict with its Arab neighbors, and Iran and has clashed on its northern border with Hezbollah—a Shi'ite Islamic movement with strong Iranian and Syrian sponsorship. Lebanon remains fragile after the Syrian withdrawal; it continues to experience bombings and assassinations against journalists and political rivalries, and its factions still present the threat of another round of civil war.

- The southern Gulf States are relatively stable and have resolved many of their border disputes in recent years, but there has been civil violence in Bahrain between Sunnis and Shi'ites. Saudi Arabia has growing problems with al-Qaeda and Islamic extremists, but the violence seems to have subsided. There are extremist elements in every southern Gulf State. Islamic extremism and terrorism are at least low-level problems in Yemen, and many southern Gulf States are heavily dependent on foreign workers, raising serious issues about their future stability.

- The struggle between the "traditionalists" and "moderates" in Iran has not developed into violent clashes. Mohammad Khatami was replaced by what many see as a hard-liner, President Mahmoud Ahmadinejad. The moderates have expressed deep dissatisfaction about their ability to accomplish meaningful reform, and there still is a serious risk of internal clashes between reformers and the supporters of Ayatollah Ali Khameni. Iran also presents major problems in terms of proliferation, becoming a nuclear power, its opposition to the Arab-Israeli peace process, its involvement in Iraq, and continued hostility to any U.S. presence in the Gulf.

- The fall of Saddam Hussein in 2003 has removed one major source of instability in the region, but "the war after the war" has seriously damaged some aspects of Iraq's oil industry and has increased the sense of instability and uncertainty in the region. There is also a risk that the ongoing insurgency may prove some form of lingering civil conflict or a long period of internal instability. Iraq is divided along sectarian lines between Sunnis and Shi'ites, and Arabs, Kurds, and Turkmen. Nation building in Iraq presents major challenges and risks. The Iraq War has already seriously cut Iraq's oil exports, and there is no current way to predict the future development of its petroleum industry and exports.

- The civil war in the Sudan has entered its second decade, and the death toll from fighting and starvation will probably exceed well over 1 million. The tensions in Darfur and the ethnic clashes have brought international scurrility without a meaningful plan to solve the issues. The low-level conflict continues to hinder the prospects of stability and inflow of meaningful foreign investment in Sudan's energy sectors.

- Yemen faces tensions between its government and key political and tribal groups in the South and has clashed with Eritrea over the control of islands in the Red Sea.

None of these tensions and conflicts pose immediate threats to the flow of MENA oil exports, but they have affected the development of energy supply in Algeria, Iran, Iraq, Libya, and Yemen, and new outbreaks of violence could occur in many MENA states with little or no warning. Most MENA states suffer from internal political, economic, and demographic problems that compound intraregional conflicts and tensions. Virtually all Middle East states have regimes with a high degree of authoritarianism—regardless of whether the ruler is called a King, Sheik, Sultan, President, General, or Ayatollah. Virtually all suffer from weak or failed economic development, high rates of population growth and a virtual youth explosion, aging and largely authoritarian regimes, and serious problems with internal stability.

The MENA region also suffers from a process of creeping proliferation that may ultimately change the nature of conflicts and the balance of power in the region. Algeria, Egypt, Iraq, Iran, Israel, Libya, Syria, and Yemen have all created missile programs and have at least conducted research into weapons of mass destruction (WMD). Israel is a major regional nuclear power and has chemical and biological programs.

Egypt has chemical and biological research programs. Iran and Syria are either trying to develop biological and chemical weapons or have already deployed them. Iran seems to have made a major effort to acquire nuclear weapons, and it remains unclear how advanced and how long it may take the Iranians to acquire a nuclear bomb. So far, such weapons have been used only in the Yemeni civil war and in the Iran-Iraq War, but there is little doubt that the Middle East is acquiring far more lethal chemical, biological, radiological, and nuclear (CBRN) weapons and delivery systems than it has possessed in the past.

The threat of Iranian proliferation is particularly important because it is centered in the Gulf region and the location of most of the world's proven oil reserves and export capacity. Both the West and the Gulf States have doubts about the stability of Iran's political leadership and the caution it will show in trying to use the threat of force to intimidate its neighbors. The United States has so far sought to prevent Iranian proliferation by diplomacy and working with its allies, but the risk of U.S. or Israeli military action against Iranian nuclear and missile facilities is real and could trigger a process of escalation and new forms of a Gulf arms race whose nature and consequences are almost impossible to predict.

Iran has matched its efforts to proliferate with the creation of significant capabilities for asymmetric warfare, such as submarines, mine warfare capability, and anti-ship missiles, and the naval branch of its Revolutionary Guards could threaten tanker traffic through the Gulf or oil and infrastructure facilities within it. Iran's recent political attacks on Israel's right to exist and denying the Holocaust create new tensions with Israel, and Iran's role in Iraq, Lebanon, and dealing with Palestinian militant groups provides further sources of tension.

MILITARISM, MILITARY EXPENDITURES, AND ARMS IMPORTS

One of the problems in analyzing trends in the MENA area is the tendency to separate the analysis of the overall trends in the economy and state spending from the trends in energy spending and from military spending and arms imports. This often suits the regimes involved, which do not want serious public or external debate over their use of energy revenues or military spending and arms purchases, and which often seek to avoid close examination of any aspect of their overall level of state spending.

As Chapter 3 discusses in detail, however, the fact remains that the past failure of many MENA states to invest in economic development, to modernize their economies, and to privatize state industry often interacts in embarrassing ways with a lack of any clear midterm to long-term energy investment strategy and their overspending on military forces and arms imports.

The level of waste in MENA military efforts, and the burden they place on economic and energy development, is difficult to put in perspective—particularly because reliable data are not always available for recent years. It has long been clear, however, that the region's militarism poses serious dangers to its peoples as well as to its stability as an energy supplier.

Several factors are involved. One is the history of crises escalating into serious conflicts, some of which have led to energy interruptions. Another is a level of total expenditure—often without a clear threat and/or producing any effective defense capability—that is so high that it seriously limits the funds available for development, including energy investment.

In addition, much of the southern Gulf countries' efforts to modernize their militaries have been focused on "glitter" purchases with little meaningful long-term and cost-effective strategic planning. Finally, many MENA countries are finding it harder and harder to sustain their present conventional force structures. At least in some cases, this adds to the pressures to acquire weapons of mass destruction and to proliferate. Such weapons present a major potential future threat to MENA energy facilities and exports.

The Overall Level of Military Effort

There has been a decline in MENA military expenditures and arms imports since the end of the Cold War. Middle Eastern military expenditures dropped from $93.0 billion in 1985 to $38.4 billion during 1997–2000.[1] Nevertheless, the region still spends nearly 6.8 percent of its gross national product (GNP) on military expenditures, and this compares with an average of only 2.3 percent for the developed world and 2.7 percent for the developing world. Militarism remains a serious problem in the MENA area. It remains the largest arms market in the developed world. It currently accounts for roughly half of the world's conventional weapons purchase agreements.[2] Once again, a quick look at the key numbers and trends that are involved can often speak louder than words:

- Figure 2.1 illustrates the scale of MENA military efforts in terms of demographics. This table also provides some of the best data currently available on the number of males

Figure 2.1   The Military Demographics of the Greater Middle East: 2005–2006

Country	Total Population	Males Reaching Military Age Each Year	Males between the Ages of			Total Eligible Males	
			15–19	20–24	25–29	Total	Medically
The Levant							
Egypt	77,505,756	802,920	3,875,287	3,875,287	3,100,230	18,347,560	15,540,234
Israel	6,276,883	105,053	251,075	251,075	251,075	2,936,041	2,468,296
Jordan	5,759,732	60,625	287,986	287,986	287,986	1,573,995	1,348,076
Lebanon	3,826,018	–	153,040	191,300	229,561	974,363	821,762
Syria	18,448,752	225,113	1,106,925	922,437	737,950	4,356,413	3,453,888
The Gulf							
Bahrain	688,345	6,013	27,533	27,533	27,533	202,126	161,372
Iran	68,017,860	862,056	4,081,071	4,761,250	340,089	18,319,545	15,665,725
Iraq	26,074,906	298,518	1,564,494	1,303,745	1,042,996	5,870,640	4,930,074
Kuwait	2,335,648	18,743	93,425	163,495	210,208	864,745	737,292
Oman	3,001,583	26,391	150,079	120,063	120,063	719,871	581,444
Qatar	863,051	7,851	34,522	34,522	43,152	302,873	238,566
Saudi Arabia	26,417,599	247,334	1,320,879	1,585,055	1,585,055	7,648,999	6,592,709
UAE	2,563,212	30,706	152,792	128,160	102,528	653,181	526,671
Yemen	20,727,063	236,517	1,243,623	1,036,353	829,082	4,058,223	2,790,705
North Africa							
Algeria	32,531,853	374,639	1,951,911	1,951,911	1,626,592	8,033,049	6,590,079
Libya	5,765,563	62,034	288,278	288,278	288,278	1,505,675	1,291,624
Morocco	32,725,847	353,377	1,636,292	1,636,292	1,309,033	7,908,864	6,484,787
Tunisia	10,074,951	108,817	503,747	503,747	503,747	2,441,741	2,035,431
Chad	9,657,069	95,228	482,853	386,282	386,282	1,559,382	834,695
Mauritania	3,086,859	–	154,342	123,474	123,474	606,463	370,513
Others							
Afghanistan	29,928,987	275,362	1,496,449	1,496,449	1,197,159	4,952,812	2,662,946
Djibouti	476,703	–	23,835	23,835	19,068	95,328	46,020
Eritrea	4,669,638	–	233,481	233,481	186,785	–	–
Ethiopia	73,053,286	803,777	4,383,197	3,652,664	2,922,131	14,568,277	8,072,755
Somali	8,591,629	–	429,581	343,665	257,748	1,787,727	1,022,360
Sudan	40,187,486	442,915	2,411,249	2,009,374	1,607,499	8,291,695	5,427,474
Turkey	69,660,559	679,734	3,483,027	3,483,027	3,483,027	16,756,323	13,905,901

Source: CIA *World Fact Book 2005* and IISS *Military Balance 2005–2006*.

entering the labor forces and in the age groups most needing jobs and most prone to terrorism and violent political action.

- Figure 2.2 summarizes the present strength of MENA military forces. It covers the "perceptual balance" between the four subregions. These data have been drawn from the IISS *Military Balance 2005–2006* and the CIA *World Fact Book 2005*.

- Figure 2.3 shows the IISS estimates of the MENA region's defense spending in $billion and as a percentage of GDP. They show the same general trend of decreasing defense expenditures relative to GDP growth.

Figure 2.2 The "Perceptual Balance": Military Forces of the Greater Middle East

Country	Total Active Manning	Total Active Army Manning	Tanks	OAFVs	Artillery	Combat Aircraft	Armed Helicopters
The Levant							
Egypt	468,500	340,000	3,855	5,682	650	572	110
Israel	168,300	125,000	3,657	10,827	1,076	402	95
Jordan	100,500	85,000	1,120	1,576	1,233	100	40
Lebanon	72,100	70,000	310	1,335	541	6	2
Syria	307,600	354,000	4,600	4,600	1,960	526	71
The Gulf							
Bahrain	11,200	8,500	180	306	69	33	24
Iran	420,000	350,000	1,693	1,285	8,196	255	50
Iraq	179,800	79,000	N/A	N/A	N/A	N/A	N/A
Kuwait	15,500	11,000	368	771	218	29	16
Oman	41,700	25,000	154	336	233	36	...
Qatar	12,400	8,500	30	334	89	12	11
Saudi Arabia	199,500	75,000	1,055	4,460	868	276	...
UAE	50,500	44,000	545	1,403	501	57	30
Yemen	66,500	60,000	790	1,040	1,167	71	89
North Africa							
Algeria	137,500	120,000	920	2,133	1,019	164	33
Libya	76,000	45,000	2,025	2,065	2,421	342	60
Morocco	200,800	180,000	656	1,219	2,892	66	19
Tunisia	35,300	27,000	132	328	276	15	...
Chad	30,350	25,000	60	203	5	...	2
Mauritania	15,870	15,000	35	95	194	...	...
Others							
Afghanistan	27,000	27,000	Some	Some	Some	5	5
Djibouti	9,850	8,000	...	21	51	...	...
Eritrea	201,750	200,000	150	80	170	13	1
Ethiopia	182,500	180,000	250	400	460	46	25
Somali	...	...	N/A	N/A	N/A	N/A	N/A
Turkey	514,850	402,000	4,205	4,543	7,450	445	37

Source: CIA *World Fact Book 2005* and IISS *Military Balance 2005–2006*.

- Figure 2.4 shows why the MENA region is described in various assessments as the most militarized area in the world. While current data are not available in a directly comparable form, the CIA *World Fact Book* and the numbers in Figure 2.3 indicate that the basic trends shown in this figure, and the other charts based on U.S. State Department, Bureau of Verification and Compliance, *World Military Expenditures and Arms Transfers, 1989–1999*, do reflect valid current trends.

- Figure 2.5 shows that the historical overall level of military effort in the region has dropped since the end of the Gulf War. This figure does not reflect the impact of the Israeli-Palestinian conflict that began in September 2000, the Iraq War in 2003, the higher spending on internal security, and the recent surge in oil revenues. Other data,

Figure 2.3 IISS Estimates of Military Spending Relative to GDP by Country: 1985–2004

Country	Military Spending (in $billion)					Military Spending as a Percentage of GDP (in %)				
	1985	1990	1991	2000	2004	1985	1990	1995	2000	2004
Algeria	0.95	**0.90**	**1.20**	**3.00**	2.87	1.94%	1.99%	2.45%	5.47%	3.71%
Bahrain	0.13	**0.18**	**0.26**	**0.32**	0.53	3.00%	4.56%	5.22%	4.67%	4.92%
Egypt	**4.14**	4.27	**2.40**	**4.40**	2.50	9.51%	10.82%	4.29%	4.50%	3.28%
Iran	**14.09**	3.18	**2.50**	**4.00**	4.41	8.62%	5.35%	4.00%	5.48%	2.98%
Iraq	**12.87**	8.61	**2.70**	**1.40**	...	57.14%	21.11%	14.75%	9.09%	...
Israel	4.23	6.16	**7.20**	**9.60**	7.87	19.18%	12.03%	9.23%	8.81%	6.73%
Jordan	0.52	0.57	**0.44**	**0.79**	0.96	12.28%	14.75%	6.67%	10.42%	8.77%
Kuwait	1.64	**13.10**	**3.10**	**3.70**	4.27	7.63%	51.76%	11.61%	11.08%	8.39%
Lebanon	0.31	0.14	**0.41**	**0.58**	0.53	12.52%	4.15%	5.29%	3.61%	3.12%
Libya	0.71	1.51	**1.40**	**1.20**	**0.62**	3.56%	5.21%	5.60%	8.89%	2.33%
Morocco	0.52	**1.21**	**1.30**	**1.40**	2.07	4.25%	4.77%	4.13%	4.24%	3.95%
Oman	2.08	1.39	**1.80**	**2.10**	3.02	21.09%	15.17%	14.75%	11.86%	12.48%
Qatar	1.66	1.08	**0.33**	**1.20**	2.19	54.41%	15.32%	4.41%	7.32%	7.77%
Saudi Arabia	22.67	**31.86**	**13.20**	**22.00**	21.30	24.21%	36.22%	10.56%	11.89%	9.03%
Sudan	0.21	**0.46**	**0.39**	**0.58**	0.48	2.89%	4.17%	4.27%	4.14%	2.49%
Syria	3.48	1.62	**2.00**	**1.50**	1.72	18.20%	9.30%	6.67%	10.95%	7.75%
Tunisia	0.42	0.40	**0.37**	**0.36**	0.44	5.06%	3.22%	2.03%	1.67%	1.54%
UAE	**2.04**	**2.59**	**1.90**	**3.00**	2.65	7.94%	7.69%	4.87%	5.17%	2.96%

Note: The IISS does not always report military expenditures (budget plus procurement costs). The numbers in bold represent defense budgets.
Source: IISS *Military Balance*, various editions.

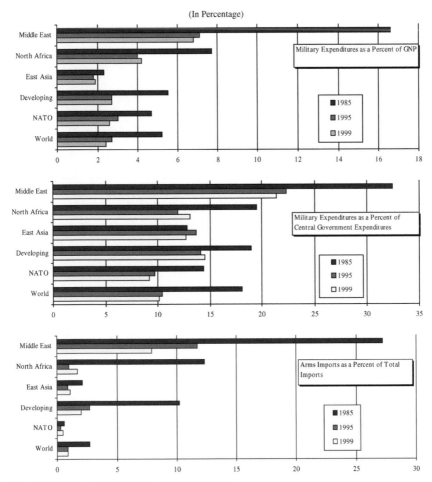

*Note: Middle East does not include North African states other than Egypt.
Source: US State Department, World Military Expenditures and Arms Transfers, 1989-1999.

Figure 2.4   "The Most Militarized Region in the World": 1985–1999

however, indicate that the decline in overall regional military efforts has continued and even accelerated in some cases. The fall of Saddam Hussein's regime in April 2003 is also likely to result in further cuts in military efforts in the Gulf area.

• Figure 2.6 shows that economic growth and overall central government expenditures are outpacing military expenditures and arms imports. These numbers are adapted from the U.S. State Department 1999 report and are likely to be different from current trends. In the period between 2004 and 2006, the MENA region has experienced high rates of economic growth due to the rise of oil prices and repatriation of large sums of capital. Countries in the region, as Chapter 3 discusses in detail, are earning large budget surpluses in 2005–2006. This growth is coupled with no major weapons purchases in recent years.

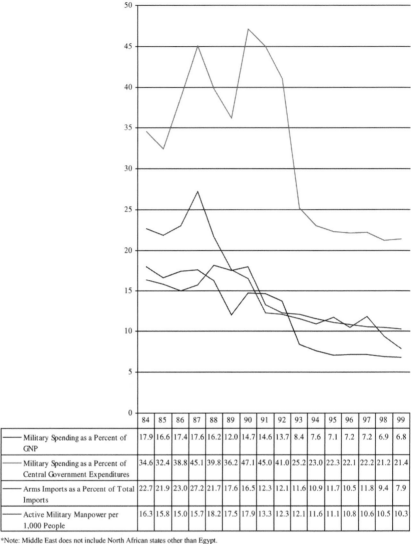

	84	85	86	87	88	89	90	91	92	93	94	95	96	97	98	99
Military Spending as a Percent of GNP	17.9	16.6	17.4	17.6	16.2	12.0	14.7	14.6	13.7	8.4	7.6	7.1	7.2	7.2	6.9	6.8
Military Spending as a Percent of Central Government Expenditures	34.6	32.4	38.8	45.1	39.8	36.2	47.1	45.0	41.0	25.2	23.0	22.3	22.1	22.2	21.2	21.4
Arms Imports as a Percent of Total Imports	22.7	21.9	23.0	27.2	21.7	17.6	16.5	12.3	12.1	11.6	10.9	11.7	10.5	11.8	9.4	7.9
Active Military Manpower per 1,000 People	16.3	15.8	15.0	15.7	18.2	17.5	17.9	13.3	12.3	12.1	11.6	11.1	10.8	10.6	10.5	10.3

*Note: Middle East does not include North African states other than Egypt.
Source: US State Department, World Military Expenditures and Arms Transfers, various editions.

Figure 2.5   Middle Eastern Military Spending as a Percentage of GNP, Total Government Expenditures, Total Population, and Arms Imports: 1984–1999

- Figure 2.7 shows that military expenditures represent far too large a portion of central government expenditures in many MENA states, limiting the funds available for development and energy investment. In some cases, military expenditures are so large that they either compound the burden excessive state spending puts on the entire economy or come close to dominating that burden as the largest single aspect of "statism." This

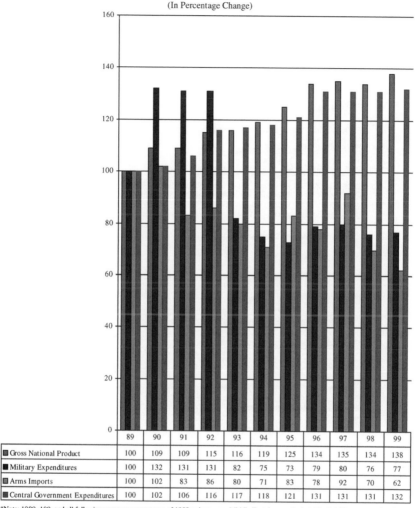

(In Percentage Change)

	89	90	91	92	93	94	95	96	97	98	99
■ Gross National Product	100	109	109	115	116	119	125	134	135	134	138
■ Military Expenditures	100	132	131	131	82	75	73	79	80	76	77
▨ Arms Imports	100	102	83	86	80	71	83	78	92	70	62
■ Central Government Expenditures	100	102	106	116	117	118	121	131	131	131	132

*Note: 1989=100, and all following years are percentages of 1989 as base year. Middle East does not include North African states other than Egypt. This was estimated using In Constant 1989 $Millions.
Source: US State Department, World Military Expenditures and Arms Transfers 1999-2000.

Figure 2.6   Middle Eastern Military Expenditures and Arms Imports Relative to Economic Growth and Government Spending: 1989–1999

combination of excessive military spending, state spending, and inefficient state industry blocks the growth and diversification of the economy as well as the work of market forces.

- Figure 2.8 shows a declassified U.S. estimate of the long-term trend in Middle Eastern (less North Africa) military spending and arms imports in current dollars. It reflects the massive swings that can occur in a time of war, but also a trend toward smaller defense expenditures and more limited arms imports.

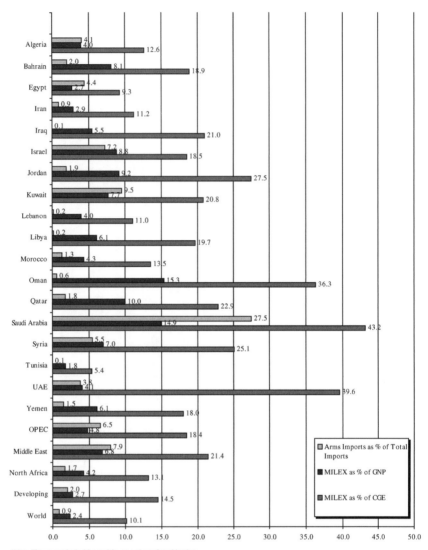

*Note: Figures marked with asterisks are estimated or older data.
Source: US State Department, World Military Expenditures and Arms Transfers 1999-2000.

Figure 2.7 Military Spending as a Percentage of Central Government Expenditures (CGE) and Gross National Product (GNP), and Arms Imports as a Percent of Total in 1999

- Figure 2.9 shows more recent trends in current dollars. The capping of military expenditures and limits to arms imports emerge more clearly. One key implication of these data, however, is that most Middle Eastern states are not spending enough to recapitalize their force structures and maintain modern forces at anything like their present equipment strength. The slope of arms imports would have to have risen steadily

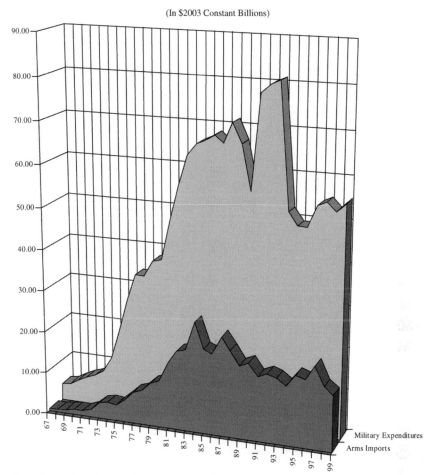

(In $2003 Constant Billions)

*Note: Middle East does not include North African states other than Egypt.
Source: US State Department, World Military Expenditures and Arms Transfers, various editions.

Figure 2.8   Trends in Middle Eastern Military Spending and Arms Transfers: 1973–1999

from 1991 onward, and spending levels would have had to nearly double in constant dollars to maintain both the force size shown in Figure 2.2 and the force quality.

• Figure 2.10 shows how these trends in North African spending relate to the overall trends in the regional economy. There has been a decline in the burden of military spending, but it is still high. The burden of arms imports is far lower, but has been heavily offset by internal security spending.

The Problem of Arms Imports

The decline in MENA arms imports is, for the most part, not a matter of choice, but a reality forced upon these countries by the need to spend more on internal

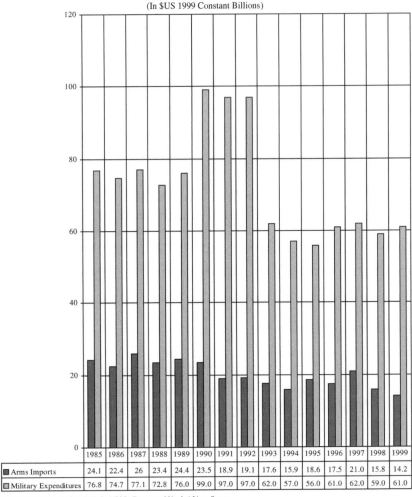

(In $US 1999 Constant Billions)

	1985	1986	1987	1988	1989	1990	1991	1992	1993	1994	1995	1996	1997	1998	1999
■ Arms Imports	24.1	22.4	26	23.4	24.4	23.5	18.9	19.1	17.6	15.9	18.6	17.5	21.0	15.8	14.2
▢ Military Expenditures	76.8	74.7	77.1	72.8	76.0	99.0	97.0	97.0	62.0	57.0	56.0	61.0	62.0	59.0	61.0

*Note: The numbers include both the Middle Eastern and North African States.
Source: US State Department, World Military Expenditures and Arms Transfers 1999-2000.

Figure 2.9   The Trend in MENA Military Expenditures and Arms Transfers: 1989–1999

security and counterterrorism, by the end of concessionary arms transfers by the FSU, by growing regional economic problems, and by a range of sanctions on key states like Iran, Iraq, and Libya.

At the same time, the drop in spending on arms imports during the 1990s was still high enough to divert important sums of money away from economic and energy development. Most MENA countries could not afford to modernize or "recapitalize" their existing force structures as they had in the past. They first tried to maintain force structures that were too large to finance, and then they bought "glitter factor" or showpiece equipment without properly finding a balanced mix of combined arms,

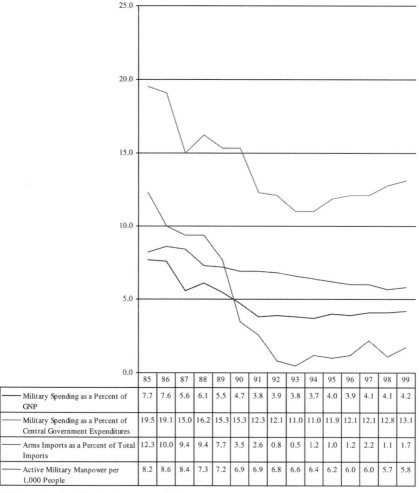

	85	86	87	88	89	90	91	92	93	94	95	96	97	98	99
Military Spending as a Percent of GNP	7.7	7.6	5.6	6.1	5.5	4.7	3.8	3.9	3.8	3.7	4.0	3.9	4.1	4.1	4.2
Military Spending as a Percent of Central Government Expenditures	19.5	19.1	15.0	16.2	15.3	15.3	12.3	12.1	11.0	11.0	11.9	12.1	12.1	12.8	13.1
Arms Imports as a Percent of Total Imports	12.3	10.0	9.4	9.4	7.7	3.5	2.6	0.8	0.5	1.2	1.0	1.2	2.2	1.1	1.7
Active Military Manpower per 1,000 People	8.2	8.6	8.4	7.3	7.2	6.9	6.9	6.8	6.6	6.4	6.2	6.0	6.0	5.7	5.8

Middle East does not include North African states other than Egypt.
Source: US State Department, World Military Expenditures and Arms Transfers, various editions.

Figure 2.10   North African Military Efforts Declined Sharply as a Percent of GNP, Government Expenditures, Imports, and Total Population: 1985–1999

or proper mix of munitions, battle management, targeting and sensor, maintenance, sustainability, and mobility capabilities. The cumulative impact was a drop in force quality so broad that it sharply limited the ability of MENA countries to fight powers such as the United States; it did not, however, affect their ability to fight each other since most countries were left at nearly the same level of relative force quality.

Recent rises in oil revenues remove many of the financial constraints placed on exporters like Saudi Arabia and ease them for states like Algeria and Libya. However, the nature of threat most MENA states now face has also changed to a mix of proliferation, asymmetric warfare, and terrorism. Many of the MENA armed forces are still struggling to account for this and to modernize their forces accordingly. Some have, and that may explain their reduced arms imports. In fact, a closer analysis of the trends in arms imports reveals why some countries are moving away from conventional forces and are trying to acquire weapons of mass destruction, long-range delivery systems, and carefully selected conventional weapons.

Arms import data are hard to find. The U.S. State Department no longer publishes its annual report on military expenditures and arms transfers, and the declassified data that the U.S. intelligence community now provides to the International Institute of Strategic Studies, and to the Congressional Research Service, cover only a few of the areas needed for in-depth analysis. Many other sources on arms transfers are little more than guesstimates and are highly unreliable. Nevertheless, there are enough data in reliable sources to determine that the general level and the direction of the trends have not changed drastically. The following figures give a broad quantitative analysis on the region's trends since the 1980s:

- Figure 2.11 puts the MENA level of arms imports in global perspective. It should be noted, however, that this table also shows that MENA arms deliveries cost some $15 billion a year, and these figures do not include the cost of military related imports of civilian and dual-use equipment like trucks, communications, and so forth. The true cost is clearly in excess of $20 billion that could otherwise be used for economic or energy development.

- Figure 2.12 shows how cumulative trends relate to the impact of given wars. Presumably, the fall of Saddam Hussein will cause a further cut in regional arms imports, although no Gulf country has yet announced its plans. It should also be noted that the data are somewhat skewed by the fact that the Israeli-Palestinian War does not involve major arms imports, and Israel's true level of imports is grossly understated in any case because only complete weapons sales—not components for military industries—are counted. Similarly, Algeria's imports relating to its civil war are often civilian or dual-use goods. The graph also sharply understates spending for nations like Iran and Syria because the costs do not include proliferation and imports for WMD.

- Figure 2.13 uses declassified U.S. data on arms sales to show the trend by major Middle Eastern country for both new arms orders and deliveries. The dominant role of two critical energy exporters—Saudi Arabia and the United Arab Emirates—is clearly apparent. The Saudi data are particularly striking because many of the nation's economic and energy development problems could be solved if it cut its arms imports to more rational levels.

- Figure 2.14 shows similar data on North African arms imports. The militaristic character of Algeria is clearly apparent. So is the impact of UN sanctions in producing a sharp decline in Libyan arms imports, effectively putting an end to Qadhafi's dreams of becoming a serious regional military power.

The Threat MENA Force Developments Pose to Energy Facilities

War and military forces have long affected the development and security of MENA energy supplies. The Arab-Israeli wars of 1956 and 1967 each affected the

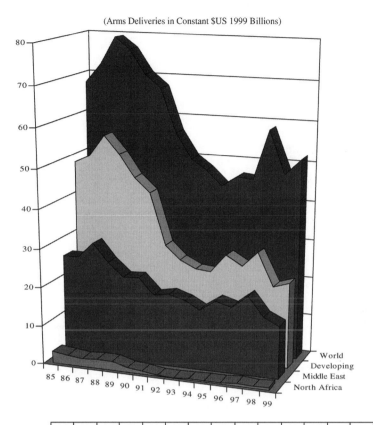

(Arms Deliveries in Constant $US 1999 Billions)

	85	86	87	88	89	90	91	92	93	94	95	96	97	98	99
■ North Africa	3.2	2.6	2.2	2.5	2.4	1.2	0.8	0.3	0.1	0.4	0.4	0.4	0.7	0.4	0.7
■ Middle East	26.2	26.2	30.2	25.6	22	22.3	18.1	18.8	17.5	15.5	18.2	17.1	20.3	15.4	13.5
▨ Developing	48.7	50.7	56	51.5	45.8	42.2	29.1	25.7	23.6	23.3	28	25.6	29.3	20.9	22.2
■ World	67.9	72.5	79.3	76.5	70.5	67.2	56.2	49.9	47.4	43.5	46.1	45.8	58.4	47.5	51.6

Middle East does not include North African states other than Egypt.
Source: US State Department, World Military Expenditures and Arms Transfers, various editions.

Figure 2.11   MENA Arms Deliveries Are Declining: 1985–1999

flow of exports to some degree, although at a time the world was far less dependent on the MENA region for its energy needs. The October War of 1973 triggered an oil embargo that led to a drastic strategic reappraisal of the importance of energy exports and imports. The Iran-Iraq War of 1980–1998, and the closely related U.S.–Iranian tanker war of 1987–1988, involved deliberate and systematic attempts to target energy production and export capabilities in a prolonged conflict. In 1991, Iraq burned Kuwait's oil fields as it withdrew from Kuwait, and the Iraq War of 2003 has witnessed both looting and sabotaging of oil pipelines, resulting in a serious reduction of Iraq's oil production and export capacity.

(Constant $US 1999 Billions)

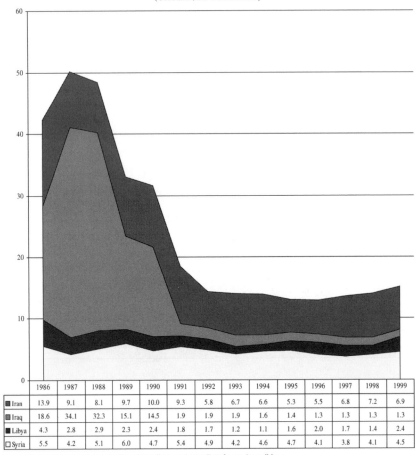

	1986	1987	1988	1989	1990	1991	1992	1993	1994	1995	1996	1997	1998	1999
■ Iran	13.9	9.1	8.1	9.7	10.0	9.3	5.8	6.7	6.6	5.3	5.5	6.8	7.2	6.9
▥ Iraq	18.6	34.1	32.3	15.1	14.5	1.9	1.9	1.9	1.6	1.4	1.3	1.3	1.3	1.3
■ Libya	4.3	2.8	2.9	2.3	2.4	1.8	1.7	1.2	1.1	1.6	2.0	1.7	1.4	2.4
☐ Syria	5.5	4.2	5.1	6.0	4.7	5.4	4.9	4.2	4.6	4.7	4.1	3.8	4.1	4.5

Source: US State Department, World Military Expenditures and Arms Transfers, various editions.

Figure 2.12   The Cumulative Decline in Military Spending by Selected Major Buyers: 1984–2003

The net impact of such wars, however, has so far been relatively limited. As yet, no country has had both the motive and the capability to launch well-planned precision strikes against an opponent's energy facilities and exports, although Iran and Iraq at least attempted to carry out such attacks during 1980–1988. The overthrow of Saddam Hussein's regime has removed one of the few governments in the MENA region that was willing to conduct such attacks without extreme provocation, and it is unclear that any other government has an incentive to conduct some form of energy war in the near future. If anything, it is terrorism, not state-vs.-state conflicts, that is likely to be the future threat.

The downward trend in MENA military forces, spending, and arms transfers does not, however, mean that MENA energy facilities and exports are necessarily safe.

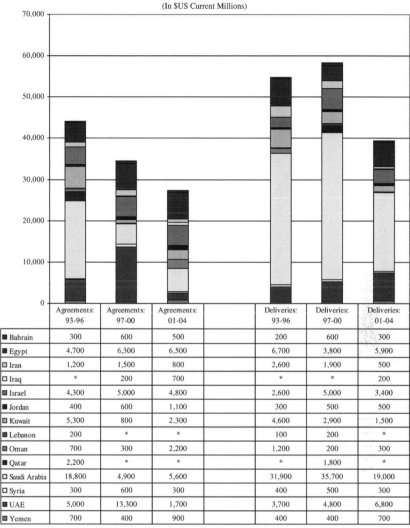

(In $US Current Millions)

	Agreements: 93-96	Agreements: 97-00	Agreements: 01-04	Deliveries: 93-96	Deliveries: 97-00	Deliveries: 01-04
■ Bahrain	300	600	500	200	600	300
■ Egypt	4,700	6,300	6,500	6,700	3,800	5,900
▢ Iran	1,200	1,500	800	2,600	1,900	500
▢ Iraq	*	200	700	*	*	200
▨ Israel	4,300	5,000	4,800	2,600	5,000	3,400
■ Jordan	400	600	1,100	300	500	500
▨ Kuwait	5,300	800	2,300	4,600	2,900	1,500
■ Lebanon	200	*	*	100	200	*
▨ Oman	700	300	2,200	1,200	200	300
■ Qatar	2,200	*	*	*	1,800	*
▢ Saudi Arabia	18,800	4,900	5,600	31,900	35,700	19,000
▢ Syria	300	600	300	400	500	300
■ UAE	5,000	13,300	1,700	3,700	4,800	6,800
■ Yemen	700	400	900	400	400	700

*Note = Data less than $50 million or nil. All data rounded to the nearest $100 million.
Source: Richard F. Grimmett, Conventional Arms Transfers to the Developing Nations. Congressional Research Service, various editions.

Figure 2.13   Trends in Middle Eastern Arms Agreements and Deliveries by Country: 1993–2004

Neither does the ongoing shift in security spending to defend against terrorist and asymmetric attacks and to improve the capability to respond to such attacks if they succeed. Most energy-exporting countries are making significant improvements in their internal security forces and have improved their ability to protect their energy facilities and critical infrastructure. But such improvements are generally still in progress and face many problems and gaps. These forces lack extensive experience,

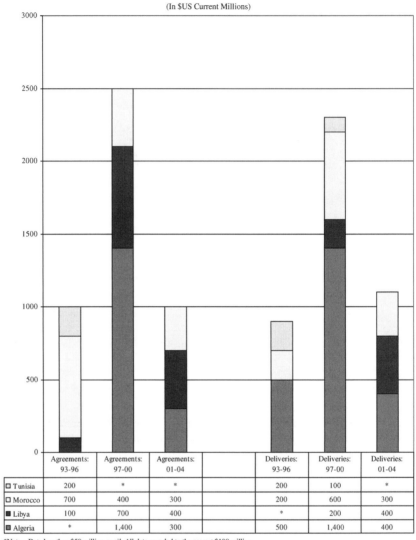

	Agreements: 93-96	Agreements: 97-00	Agreements: 01-04		Deliveries: 93-96	Deliveries: 97-00	Deliveries: 01-04
☐ Tunisia	200	*	*		200	100	*
☐ Morocco	700	400	300		200	600	300
■ Libya	100	700	400		*	200	400
▣ Algeria	*	1,400	300		500	1,400	400

*Note = Data less than $50 million or nil. All data rounded to the nearest $100 million.
Source: Richard F. Grimmett, Conventional Arms Transfers to the Developing Nations, Congressional Research Service, various editions.

Figure 2.14   Trends in North African Arms Agreements and Deliveries by Country: 1993–2004

which makes them more vulnerable; and far too little has been done to reduce the physical vulnerability of facilities, provide for rapid repair, and stockpile critical components and long-lead replacements. Effective strategic reserves generally do not exist.

Both MENA nations and extremist movements are learning a great deal more about asymmetric war with time, and states are using their remaining funds to acquire more precision weapons and better platforms to launch them from. These weapons include air-launched weapons that can be used against such key targets as export facilities, major energy processing and distribution facilities and as-oil separators, desalination and war injection facilities, power plants, gas train refineries, and petrochemical plants. MENA countries are also buying better maritime surveillance systems, and longer-range antiship missiles, better mines, and submarines also allow MENA states to do a better job of attacking tankers and offshore facilities.

Putting state capabilities aside, it is terrorists that pose a new and significant threat by their sudden eagerness to attack energy facilities and by their ability to use religion to subvert security forces and those who work in such facilities. Until recently, Islamist extremist groups have either avoided energy targets or have not made them a critical priority.

This changed dramatically when the insurgency became serious in Iraq, however, and since that time key Islamist extremist leaders like Osama bin Laden have threatened such attacks. In a tape that was posted on an extremist Web site, bin Laden asserted, "Targeting America in Iraq in terms of economy and loss of life is a golden and unique opportunity...Be active and prevent them from reaching the oil, and mount your operations accordingly, particularly in Iraq and the Gulf."[3] Bin Laden's deputy, Ayman Al-Zawahiri, also urged similar attacks. On December 7, 2005, a statement attributed to Al-Zawahiri called on the "mujahideen to concentrate their attacks on Muslims' stolen oil, from which most of the revenues go to the enemies of Islam while most of what they leave is seized by the thieves who rule our countries."[4] Abu Musab Al-Zarqawi, the leader of al-Qaeda in Iraq, made similar statements urging attacks against energy facilities in the Gulf.

Iran presents a particular problem in terms of its asymmetric warfare capabilities. It occupies strategic positions on both sides of the Strait of Hormuz and in islands near the main shipping paths in the Gulf. These include bases and facilities at Bandar-e-Abbas, Khorramshahr, Larak, Abu Musa, Al Farisiyah, Halul, and Sirri. Its forces are organized and regularly train for asymmetric attacks, including attacks on large ships and facilities similar to energy and infrastructure facilities. Its 125,000-man Islamic Revolutionary Guards Corps has a 20,000–23,000-man naval force, with some 5,000 marines. While its air force is largely obsolete, it does have relatively long-range C-801K (CSS-N-4) antiship missiles.

Although its regular navy is also obsolete, virtually any Iranian ship can be used to lay floating mines, and Iran may have modern influence and bottom mines. It has three Kilo Type 877 conventional submarines with long-range homing torpedoes and mine warfare capability. The naval force in the Guards has HY-2 (CSS-C-3) antiship cruise missiles, and large numbers of small craft, some equipped with anti-tank guided missiles, rocket launchers, and recoilless rifles and machine guns. These small craft (largely Boghammer Marins) are difficult to detect by radar and visually. The naval force also has 10 small Houdong missile patrol craft with C-802 (Css-N-8) antiship missiles.

Iran may also use terrorists and extremists as proxies or as false flags for attacks on other states. Such attacks do not have to come in forms that suit Iran's ideology.

There are some indications that Iran may have already used Hezbollah in Lebanon to provide more sophisticated weapons to Sunni insurgents in Iraq in order for them to fight the United States. Syria too has tolerated insurgent and terrorist activity within its borders in Syria that supports Sunni insurgents in Iraq.

The history of war is the history of sudden explosive crises and unplanned escalation. There is no way to predict whether one MENA state will launch attacks on another's energy facilities or whether such attacks will be well planned and well executed. What is important to understand, however, is that the broad trends in MENA military forces do not necessarily improve the security of energy facilities. In fact, it is possible that MENA countries with limited conventional capabilities would lash out at high-value targets to try to defeat, intimidate, or punish their neighbors. If so, they will continue to acquire the means to conduct such attacks over time.

PROLIFERATION

Proliferation is a serious problem in the Middle East, and one that is not likely to diminish in the near future. There is a complex pattern of proliferation in the region, and the range of delivery systems is steadily expanding. Algeria, Egypt, Iran, Iraq, Israel, Libya, the Sudan, and Yemen have all been involved in past efforts to acquire weapons of mass destruction—albeit at very different levels and with different goals and intentions. Algeria, Iraq, the Sudan, and Yemen are no longer part of the list of serious proliferators, but Iran, Israel, and Syria are.

Israel has large nuclear forces, and Iran has a rapidly maturing nuclear program. Iran and Syria have significant biological warfare programs, and Egypt and Israel have conducted significant research and development activity. All of the proliferators in the Middle East are working on—or have—chemical weapons; Iran, Libya, and Syria probably have a stockpile of such weapons. Algeria, Egypt, and Israel have the technical capability to produce them. Egypt, Iran, Iraq, Israel, Libya, and Saudi Arabia have long-range missiles or programs to acquire them.

Terrorist movements such as al-Qaeda have also sought weapons of mass destruction, and there is no way to predict how asymmetric wars or terrorism using CBRN weapons could damage oil and gas production and petroleum export facilities or the level and duration of any interruption in exports. It is all too clear, however, that the threat posed by future wars and terrorism is becoming far more serious than in the past.

Proliferators in the Middle East

Figure 2.15 shows the list of MENA proliferators and the current estimated status of their efforts. The nations listed in Figure 2.15 are so different in terms of regime, goals, and behavior that it is obvious that there is no regional threat to the West, but rather the possibility that individual states might pose a threat to individual Western nations or interests. Two major proliferators—Iran and Libya—are of special interest. These are nations that have posed a threat to the West in the past and which have

Figure 2.15   Nations with Weapons of Mass Destruction

Country	Type of Weapon of Mass Destruction		
	Chemical	Biological	Nuclear
East-West			
Britain	Breakout	Breakout	Deployed
France	Breakout	Breakout	Deployed
Germany	Breakout	Breakout	Technology
Sweden	–	–	Technology
Russia	Residual	Residual	Deployed
United States	Residual	Breakout	Deployed
Middle East			
Algeria	Technology	Technology	Interest
Egypt	Residual	Breakout	–
Israel	Breakout	Breakout	Deployed
Iran	Deployed?	Breakout	Technology
Iraq	?	?	Technology
Libya	Deployed	Research	–
Syria	Deployed	Technology?	–
Yemen	Residual	–	–
Asia and South Asia			
China	Deployed?	Breakout?	Deployed
India	Breakout?	Breakout?	Deployed
Japan	Breakout	Breakout	Technology
Pakistan	Breakout?	Breakout?	Deployed
North Korea	Deployed	Deployed	Deployed (?)
South Korea	Breakout?	Breakout	Technology
Taiwan	Breakout?	Breakout	Technology
Thailand	Residual	–	–
Vietnam	Residual	–	–
Other			
Argentina	–	–	Technology
Brazil	–	–	Technology
South Africa	–	–	Technology

also sponsored attacks of state terrorism against Western targets and/or on Western soil.

Iran currently poses the most significant near-term threat in terms of acquiring biological and nuclear weapons, and long-range missiles that might strike Europe or the United States. In spite of Iranian denials, there is little doubt that Iran has an active nuclear and biological weapons program and has already begun to test long-range missiles. Iran's capabilities, however, will remain uncertain. It also faces a strong regional threat from Israel, and diplomatic pressures and threats of sanctions by the European Union and the United States. While Iran's regime may or may not become truly moderate in character, it has become more pragmatic since the death of Ayatollah Ruhollah Khomeini, and it is far from clear that it would take "existential" risks of the kind posed by such an attack on the West.

There has, however, been growing international pressure on Iran to cut back on its nuclear program. The discovery of two major undeclared underground facilities in 2003 that included a centrifuge plant suitable for producing fissile uranium, and a heavy water plant for a reactor fuel cycle that could produce weapons grade pluto-nium, led to broad international pressure on Iran to permit full-scale inspection on the terms provided to the protocol for the Nuclear Non-Proliferation Treaty.

So did a follow-up analysis in 2004 and 2005 by the International Atomic Energy Agency (IAEA) that investigated Iran's nuclear program in depth and reported a number of forbidden research and development activities and confirmed the fact that Pakistan had secretly sold nuclear weapons technology and design to Iran. Iran is also proceeding with long-range missile developments that have little military meaning unless the missiles are armed with weapons of mass destruction, and there are no effective limits on its chemical and biological weapons programs.

The Gulf Cooperation Council (GCC) States, their energy facilities, and critical infrastructure like desalination plants and electric power plants are the closest to Iran's nuclear program. The Gulf States have tried to avoid open confrontation with Iran, but have expressed concern about Iran's nuclear program. For example, in the GCC summit in December 2005, the GCC Secretary General, Abdul Rahman Al-Attiya, expressed the GCC view over Iran's nuclear reactors by saying, "If it is not for peaceful applications, then the program becomes unjustified and the issue cannot be neglected...but we do not see Iranian nuclear reactor as a cause of danger and instability to the region....We expect Iran to be rational in dealing with the nuclear issue, that it meets peaceful purposes without inflicting damage on its neighbors."[5]

Libya has the dubious distinction of being the only MENA state to have fired a long-range missile on a Western target—it fired on the Italian island of Lampadusa following the U.S. raid on Tripoli. At the same time, a lot of Libya's grandiose mili-tary plans have so far ended in failure. Libya has some chemical weapons capability, but has failed to develop ballistic missiles with longer ranges than the Scud. It has explored biological and nuclear weapons programs, but there is little evidence of suc-cess. Moreover, the interception of a ship carrying nuclear centrifuge and weapons technology to Libya in the fall of 2003 led to political pressure that resulted in Lib-ya's declaration in December 2003 it would give up all of its programs for developing missiles and weapons of mass destruction and allow unconditional inspections. This was followed by U.S., U.K., and IAEA inspections and the transfer of much of Lib-ya's technology to the United States.

Given this background, it currently seems unlikely that even the most radical MENA power would readily take the risk of directly confronting a combination of its neighbors, the United States, and other Western states, given the relative military weakness of key potential threats and the risk of massive retaliation. No Middle East-ern state can disregard the fact that any use of a biological or nuclear weapon that produces massive casualties could trigger devastating conventional strategic strikes or even the use of nuclear weapons by the West.

At the same time, terrorist and extremist movements are far harder to deter than states. The history of the region is filled with miscalculations, erratic behavior, and risk taking. Behavior can alter rapidly in a crisis, and the most threatening states have rulers or ruling elites that may choose to escalate in ways that are far less conservative than what Western planners would consider under similar conditions. Extremist movements like al-Qaeda have also shown that they will take extreme risks to take extreme action, and they may either cooperate with proliferating states or act as their covert proxy.

The following scenarios could involve either the more radical proliferators in the region like Iran, or terrorist attacks. They may not represent even moderate probability cases, but they are possible enough so they deserve serious consideration:

- Weapons of mass destruction might be used against key energy and energy export facilities in intraregional conflicts to pose a major economic threat to a regime or put pressure on the West.

- Attacks might be carried out on Western power projection forces in the region, or the threat of such attacks might be used to try to force a regional power to expel Western power projection forces or to deprive regimes of Western support.

- Threats against the West, demonstrative long-range missile attacks against targets in the West, or low-level use of weapons of mass destruction might be used against targets in the region or the West to try to force Western nations to support the policies of a given Middle Eastern state or intervene in a regional conflict. The escalation of an Israeli-Syrian conflict, or future Iranian-GCC conflict, might lead to such a threat.

- A regional power might set up a launch on warning or launch-under-attack system targeted on the West in an effort to deter Western intervention or military action. Such a system might be created to prevent Western counterproliferation strikes.

- The threat, demonstrative use, or larger-scale use of such weapons might be utilized in an effort to force an end to sanctions or containment.

- A regime on the edge of collapse might lash out, feeling it had nothing to lose and accepting the risk of broader retaliation against the nation. Alternatively, a nation under nuclear attack by Israel might feel that attacks were justified against Western targets, particularly U.S. bases.

- Terrorists could use such weapons in the West to try to further divide the West and the Arab world, building on the tensions caused by the Second *Intifadha,* hostility growing out of September 11, 2001, and the Iraq War.

- Middle Eastern states are not limited to conventional forms of warfare. While a great deal of attention focuses on long-range missiles, a Middle Eastern state might use unconventional delivery means or a terrorist proxy to deliver such weapons—hoping that it would not be identified as the source or that enough ambiguity would exist to prevent a decisive response.

- Technology or fissile material transfers might suddenly destabilize the balance. This might include the transfer of long-range missiles or fissile material, or key components and technology for missiles and weapons. This could suddenly alter the regional balance and the perceived risk in threatening the West or Western interests.

- Political miscalculation by a regime in Iran, Libya, or Syria can escalate toward the use of WMD when they feel threatened by external powers.

- If insurgents in Iraq get their hands on CRBN weapons, they can use them against the U.S. forces, the Iraqi security services, or Iraqi civilians.

Once again, the problem is to balance possible risks against probable risks over a period as long as 2006–2030, knowing that new proliferators can emerge and many regional powers could acquire missiles, cruise missiles, or better strike aircraft. Most MENA states and leaders are normally cautious and self-preservation is normally the highest single priority. There is no question, however, that a combination of creeping proliferation, and having to rely on the judgment and stability of at least five major proliferators, presents risks.

The Threat Proliferation Poses to Energy Facilities

Proliferation represents the most serious potential threat to MENA energy facilities in terms of lethality. Even a small nuclear weapon could destroy any energy production or export complex in the Middle East, although oil and gas fields are too dispersed and "hardened" for such attacks to have much effect. Radiological weapons are far harder to produce in effective lethal form than many analysts seem to understand, but are probably within the state of the art of the more advanced Middle Eastern economies, and even small, relatively nonlethal, radiological weapons might keep workers from entering a site or facility.

Biological weapons can be more lethal than nuclear weapons and have the advantage of leaving facilities intact. Persistent biological weapons are possible and can be used to contaminate facilities in much the same way as radiological weapons. Once again, their psychological effect might be more important than their lethality or killing effect.

Large amounts of chemical weapons are needed to achieve high lethality against large energy facilities, but such attacks are within the state of the art for countries like Syria and Iran—which seem to have cluster munitions and persistent nerve gas. Advances in warhead and weapons design are also improving the capability to disseminate both chemical and biological weapons from missiles, cruise missiles, bombs, and unmanned combat air vehicles. There is no way to know how advanced MENA countries now are or how much progress they will make in the future, but it is clear that some could have a limited capability now and that they could all potentially acquire relatively sophisticated capabilities during 2005–2010.

Proliferating MENA nations may never use weapons of mass destruction against energy facilities, but there are incentives that could lead to making such threats or to actually using such weapons in a crisis. Energy facilities are a natural hostage in a strategic confrontation and involve far less provocation than attacks on civilian populations. The threat of such attacks might offset a major conventional advantage on the part of an opponent or be used to deter military action or the support/basing of outside powers like the United States. The effective destruction of a key energy facility could also produce a serious mid- to long-term blow in economic terms.

In any case, it is dangerous to assume that crises and escalation are handled in rational terms or from a common set of perceptions. Proliferating nations know that even the threat of such strikes can have a powerful deterrent or intimidating impact, and threats can lead to use. Moreover, world energy markets might well panic at the very threat of the use of WMD and might take days or weeks to stabilize after even a token use of a weapon whose effects would be difficult to estimate and understand.

Independent or proxy action by terrorist groups presents a further problem, as does the risk of covert attack. Extremist groups cannot present the same range of military threats as states, but this might not matter if they picked the right target. State use of a terrorist group as a proxy, or covert attack, could achieve significant results in some scenarios with far less fear of retaliation or some kind of serious action by the international community.

TERRORISM AND STATE TERRORISM

As has already been touched upon, terrorism and extremist violence are changing the strategic map of the MENA region. This threat may take the form of alliances with state actors, but most of the threat comes from independent extremist groups that wish to enforce their own values and interpretation of Islam on MENA states and other Muslims. As Chapter 3 discusses, they exploit the fact that many moderate regimes face problems with internal and external terrorism, much of it directed often against progressive change and reform and their alliance with the West.

Violence by extremist groups has already proved to be exceptionally dangerous and destabilizing. Every nation in the Middle East, no matter how moderate, faces some level of internal and external threat from such movements. Active internal fighting has taken place in Algeria, Libya, Egypt, Lebanon, Syria, Saudi Arabia, and Yemen. Iran is torn between "hard-liners" and "moderates," and the fall of Saddam Hussein has unleashed new extremist forces in Iraq.

Every other MENA country has had to establish new security procedures and cope with its own extremists groups. The problem is also an international one that reaches far outside the MENA area. It now involves Central Asia, South Asia, the Islamic countries of Southeast Asia, and movements in Europe and North America.

While militarism and proliferation pose potential threats to the region's development and energy exports, the most active threat of violence now comes from this violent extremism. It does not, however, have one source or represent one cause. Some have arisen in response to state terrorism, in response to regional conflicts like the Israeli-Palestinian War, but other elements have developed in part due to the pressures of social and economic change. The end result is a complex mix of threats that combines national movements, regional movements, and truly international movements like al-Qaeda.

The ideology and goals of these movements differ from group to group, but there are often loose alliances of groups with different goals. What most do have in common is that their ideology is based on an extremist version of Shi'ite, Sunni, Sufi, and neo-Salafi Islam and that the religious goals of each movement are mixed with a political agenda that is based on the rejection of modern economic priorities and reform. Often it is also based on a struggle for power in their respective countries and as a backlash against outside "intervention" in domestic affairs. So far, they are all small extremist groups that do not represent the views and hopes of the vast majority of the people in their country or the MENA region, but several have already proven to be dangerous both inside and outside the Middle East.

The Regional and Global Impact of Extremists Groups

Long before 9/11, the attacks on Al Khobar, the USS *Cole,* and the World Trade Center showed that terrorism posed a threat to the moderate regimes in the Middle East and a transnational threat to the West. There have been many serious terrorist attacks on Western targets in the Middle East in the past, such as the bombing of the U.S. Marine Corps Barracks in Beirut.

The November 13, 1995, truck bombing of the National Guard Headquarters in Riyadh killed five U.S. servicemen and two Iranians. The June 25, 1996, bombing of the Khobar Towers killed 19 U.S. servicemen. The attacks on the U.S. embassies in Kenya and Tanzania involved large numbers of innocent casualties—247 dead and over 5,000 wounded in the case of Kenya, and 10 dead and more than 75 wounded in the case of Tanzania. These attacks involved truck bombs with 600–800 pounds of explosives.

Civil tension in the Middle East has made tourists a target. For example, the worst terrorist attack in Egypt's history occurred on November 17, 1997. Six gunmen belonging to the Egyptian group al-Gama'at al-Islamiyya (Islamic Group or IG) entered the Hatsheput Temple in Luxor. For nearly half an hour, they methodically shot and knifed tourists trapped inside the Temple's alcoves. Fifty-eight foreign tourists were murdered, along with three Egyptian police officers and one Egyptian tour guide. The gunmen then fled the scene, although Egyptian security forces pursued them and all six were killed. Terrorists launched a grenade attack on a tour bus parked in front of the Egyptian National Museum of Antiquities in Cairo on September 18, 1997, killing nine German tourists, an Egyptian bus driver, and wounding eight others.

The attack that truly globalized Middle Eastern terrorism was the series of attacks on the World Trade Center and the Pentagon on September 11, 2001. There have been many previous attempts at such attacks, and many smaller successful attacks on targets in Europe. It was "9/11," however, that showed the United States that its territory and civil population could be as vulnerable as the nations of the Middle East.

Since the 9/11 attacks, however, Middle Eastern states have experienced a wave of attacks by groups such as al-Qaeda. For example, Morocco has also experienced attacks by al-Qaeda. On May 16, 2005, al-Qaeda struck Casablanca with five blasts (10 suicide bombers) "targeting a Jewish community centre, a Spanish restaurant and social club, a hotel and the Belgian consulate." The bombing killed 41 and injured at least 100, according to Moroccan officials.[6]

So far, the terrorist and extremist elements in the Israeli-Palestinian conflict have not attacked other states, or threatened energy facilities. The situation is very different in the Gulf.

Since al-Qaeda in the Peninsula struck at compounds of foreign workers on May 12, 2003, Saudi Arabia has been actively fighting al-Qaeda. Extremists have attacked or attempted to attack major government infrastructure, the expatriate community, and foreign interests in the Kingdom. The attacks have caused substantial damage and have added a sense of insecurity.

Prince Nayef, the Minister of Interior, summed up the damage done by such attacks in his speech at the Saudi Counter-Terrorism Conference in Riyadh in February 2005:

> In the last two years, Saudi Arabia has witnessed 22 criminal incidents—including explosions, attacks, and kidnapping—causing the death of 90 citizens and foreign nationals and injuring 507 people. Thirty-nine security troops were martyred and 213 among them were injured, whereas 92 terrorists of this miscreant minority were killed and 17 of them wounded. Material losses in property and damage to facilities have exceeded 1 billion dollars. It is thanks to Allah's grace and their alertness that the security forces have been able to foil a total of 52 terrorist operations in preemptive strikes that have thwarted the occurrence of any further loss in life or property.

It is too early to draw general conclusions about the final success of Saudi counter-terrorism efforts, but early indications show that the Saudi security forces have been on the offensive, and al-Qaeda is on playing defense. During 2005, there have been no bombings, but many shootouts between the Saudi security forces and terrorists, and more losses on both sides.

On February 24, 2006, Saudi officials reported that they had foiled a suicide attack by two cars on the oil production facility of Abqaiq, in the first effort to target the Saudi oil infrastructure directly. This event underscored both the threat posed by terrorism as well as the resilience, at least until now, of the Saudi security forces to protect the country's critical infrastructure from attack.[7]

The major uncertainty remains the future of the insurgency in Iraq, the role of the Saudi fighters in Iraq, and their future role in causing instability within the Kingdom.

Iraq's other neighbors have also been under pressure from such extremists and terrorists, and from the instability in Iraq. On November 9, 2005, three nearly simultaneous suicide bombings against hotels in Amman, Jordan, killed at least 56 people and injured some 100. Abu Musab Al-Zarqawi's group claimed responsibility for the attacks. Shortly after the attacks, the Jordanian government declared war against extremism. King Abdullah II specified the enemy as the Takfiri groups such as al-Qaeda and other splinter groups; he said that Jordan will not slow its reform efforts, and he added, "At the same time, it reaffirms our need to adopt a comprehensive strategy to confront the Takfiri culture."[8]

European cities such as Madrid and London have also seen terrorist attacks, as well as targets in East Asia. Many have attributed these attacks to the anger toward the United States and its allies for the Iraq War, given that they all happened after the start of the Iraq War in March 2003. Other experts, however, have argued that these bombings were planned before, and the Iraq War was exploited by extremists to gain popular support for their actions.

While the ongoing Iraq War has contributed to the anger in the Arab streets, the terrorist threat is multifaceted and has a complex history. The main targets for groups such as al-Qaeda are the governments in the MENA region. They are seen by extremists as puppets of the West, corrupt, and secular. While al-Qaeda emerged as the most important current threat, there were many causes of transnational terrorism in the Middle East, and there are many different targets:

• The United States is a major target because it projects the most power into the region, because of its close ties to Israel, because attacks on the United States produce the most worldwide publicity, and because the United States can often be used as a proxy for less popular attacks on Middle Eastern regimes.

• The breakdown in the Arab-Israeli peace process has triggered a wave of Palestinian attacks in response to steadily escalating Israeli "excessive force." It is a tragedy that could trigger a broader Arab-Israeli conflict and make Americans a target, both out of frustration and in an effort to break up the peace process.

• The failures of Middle Eastern governments, state terrorism and authoritarianism, economic hardship, social dislocation, and the alienation of youth combine to create extremist groups that not only attack their governments, but use Western targets as proxies. Motives can include attempting to drive out the Western military forces that provide Middle Eastern countries with security, cripple the economy to weaken governments, or win public recognition in the region. While some of these groups are secular, most are Islamic in character.

• European nations have become the scene of attacks by terrorist groups as well as opposition groups on the embassies of Middle Eastern regimes, or by opposition groups attacking each other. Iran has sponsored state terrorist attacks on the People's Mujahideen and Kurdish opposition groups in France, Germany, Switzerland, and Turkey. Israel has killed Palestinians in nations like Norway.

• Western tourists and businessmen can be the targets of terrorists in the Middle East, as such groups seek to put economic pressure on local regimes or prove their status and power.

The goals of neo-Salafi Sunni extremist groups like al-Qaeda differ in detail, but have the same major objectives. Such movements want control over Gulf and other Islamic states to use as launching grounds to expand their ideology and political influence over the rest of the MENA region and possibly the world.

Their main objectives have so far been to destabilize governments, but they have targeted civilians, foreign workers, nongovernmental organizations, critical facilities, and a host of other targets of convenience. They already have made energy facilities in Iraq a major target, and their statements indicate they now plan to broaden such attacks to other countries. This could pose particularly serious problems in the Gulf. Experts believe that since al-Qaeda has had little success in destabilizing Saudi

Arabia, and while the Saudi security apparatus has proven itself capable, it remains unclear whether the security forces in the other Gulf States can handle the same magnitude of threat and attacks. The following are key areas of uncertainties that the Gulf States need to develop their capacity to deal with:

- Countries such as the United Arab Emirates have many high-value targets such as high-rise buildings where terrorists can inflict grave damages to citizens, expatriates, and the economy.

- Security forces in Bahrain, Qatar, and the United Arab Emirates have a "non-national" component from various countries. This presents a threat for extremists to infiltrate.

- It is estimated that 81 percent of the UAE population are noncitizens, 60 percent of Qatar, 59 percent of Kuwait, 37 percent of Bahrain, 26 percent of Oman, and 24 percent of Saudi Arabia. Attacks on expatriates can cause an exodus of foreign labor and be harmful to their economies.

- There are limited data on how the GCC States protect their energy infrastructure or build redundant installations in case of damage. An attack on an oil installation in the Gulf can send oil prices even higher in an already tight energy market.

- The radicalization of the region's youth and the fear of the Iraq insurgency spilling over into neighboring states.

- Implementing meaningful cooperation on border and coast security to prevent the flow of terrorist, arms, and explosives.

Much now depends on how the Gulf States and other MENA states react. The current period of high oil prices, large oil revenues, and robust economic conditions presents an opportunity for all the Gulf States to assess their threats and craft comprehensive plans for their strategic future. This involves not only strengthening their security apparatuses, but also implementing realistic reforms in their economies, stock markets, and social structures.

In general, MENA states have so far done far more to talk about cooperating in counterterrorism than they have done to act. This is just as true of the GCC states as the other states in the region, but there have been some indications by the GCC states that they are increasing counterterrorism cooperation. In October 2005, the GCC signed an antiterror pact, and according to the Saudi Information Minister Iyad Madani, "The signatories of the agreement will cooperate with one another by providing necessary security support to any member country that faces danger or terrorism."[9]

Intelligence and security cooperation at the international and regional levels have always been controversial. There is, however, a slowly growing realization that the terrorist threat is unique and requires "collective response." Unlike more conventional forms of terrorism, such attacks deliberately seek to create a "clash of civilizations" as well as clash between civilizations. The extremists' goal is to build on other regional problems and tensions to divide the West and Arab worlds, and they have exploited many of the regional tensions to advance their political ideology.

The Clash within a Civilization, the Arab-Israeli Conflict, and the Western Counterreaction

Al-Qaeda's strategy to attempt to destabilize the Gulf and the MENA region at large is likely to continue, and one may see more bloody attacks in the Gulf and Iraq. In addition, extremist movements exist at some level in every MENA state. Extremism and terrorism remain a major threat to MENA governments, and the end result is more a clash within the Islamic civilization and not a clash between Islam and the Arab world and the West.

The primary goal of most Islamic extremist movements is not to attack the West per se, but to destabilize Islamic regimes, and to recreate the region based on their "deviant ideology." This ideology is based on radical socialism or economic change and conservative social customs. They are against any social, economic, or political modernization. Most of these extremists groups know what they are against, but they have only vague and impractical ideas of what they are for.

Some groups are against anyone who disagrees with their ideology. Some are called neo-Salafists or *takfiries* (or those who tend to excommunicate other Muslims and people of the Book and call them unbelievers), who see the world as them vs. everyone else. They have called major Sunnis, Shi'ites, and Suffi scholars apostates and they are calling for the overthrow of moderate governments in the region. This branch is represented by Abu Musab Al-Zarqawi and Osama bin Laden.

There are, however, other forms of terrorism and extremist violence. The fact that the Arab-Israel peace process has given way to an Israeli-Palestinian war has led to a new wave of violence on both sides. The Israeli side has used conventional forces to occupy and attack the Palestinians. The Palestinians have used asymmetric and guerrilla warfare—most notably in the form of suicide bombings. The Palestinian attacks on Israeli civilians have been overwhelmingly by Islamist groups like Hamas and the Palestinian Islamic Jihad, but have increasingly involved support from the hard-line elements of secular Palestinian groups as well.

The lines between Islamic extremism and the Arab-Israeli conflict have been further blurred by the role Shi'ite groups like Hezbollah (the first group to use suicide attacks against Israel) played in driving Israel out of Lebanon, and the role Iran and Syria have played in supporting Hezbollah. Syria at least tolerates militant groups on its soil that oppose Israel. Iran has increasingly funded non-Shi'ite groups like Hamas and the Palestine Islamic Jihad, and private money has flowed to such groups from the Arab states—partly to support their charities and partly to support the groups in attacking Israel.

Unlike most forms of Islamic extremism and terrorism, the Israeli-Palestinian conflict also polarizes the Arab world at a popular level. If the Israeli image is one of Palestinian terrorism, the Arab image is one of excessive Israeli use of force, continued occupation, and continued settlements. It allows extremist and terrorist groups to exploit the conflict to win popular support and to exploit the image of the United States as Israel's ally and supporter. More generally, it allows them to exploit the image of the West as exploiting the Arab world and standing with Israel against the stateless Palestinians.

The Role of the West in the Clash

The West, and particularly the United States, has often reacted by confusing Islamist extremism and terrorism with Islam, the Arab world, and Iran. U.S. officials have tried to avoid such stereotypes and dangerous generalizations, but many American and Western media and analysts have not. One of the ironies of 9/11 is that Osama bin Laden and al-Qaeda have succeeded in part in producing a Western counterreaction that does to some extent reflect a "clash between civilizations." The U.S. and British invasion and the occupation of Iraq have increased such tensions, as have the failures to bring effective security and development to Afghanistan and U.S. talk of broad regime change along lines where its concept of future "democracies" is as vaguely defined as of the future desired by most Islamist extremists.

The United States and Europe, however, need patience, a balanced approach to reform, strong country missions capable of encouraging local governments and reformers, and the understanding that different societies and cultures will often take a different path. In practice, this means a very different strategy based on persuasion, partnership, and cooperation rather than on pressure and conversion:

- **Implement a broadly based reform strategy.** Social, economic, and political reforms should be supported, but in an evolutionary sense. The United States and Western states, however, cannot be seen as pushing these reforms in ways that discredit local officials and reformers. Outside pressure for change will be resisted even if the reforms are necessary, and too much overt pressure is counterproductive.

- **Recognize that one size does not fit all.** The Arab and Islamic worlds are not monolithic. Each country requires different sets of reforms and needs. Some need help in reforming their political process, others need economic aid, and others need special attention to their demographic dynamics and population control. The West, therefore, must avoid any generalized strategy of dealing with the Arab-Islamic world as one entity.

- **Work on a country-by-country approach and rely on strong country teams, not regional approaches.** Regional policies, meetings, and slogans will not deal with real-world needs or provide the kind of dialogue with local officials and reformers, tailored pressure and aid, and country plans and policies that are needed. Strong country teams both in Washington and in U.S. embassies are the keys to success.

- **Recognize that the pace of reform will be relatively slow if it is to be stable and evolutionary, and it is dependent on partnership and cooperation.** Artificial deadlines and false crises can lead only to failed tactics and strategies. Outside support for reform must move at the pace countries can actually absorb, and priorities must be shifted to reflect the options that are actually available. History takes time and does not conform to the tenure of any given set of policy makers.

- **Carefully support moderate voices.** "Moderates" in the region do need the support of the West, but obvious outside backing can hurt internal reform efforts. Moreover, "moderate" must be defined in broad terms. It does not mean "secularist," and it does not necessarily mean "pro-American." It also, however, does not mean supporting voices that claim to support freedom and democracy, but are actually the voice of extremism.

- **Recognize that democratization is only a part of reform and depends on creating a rule of law, checks and balances and a separation of powers, protection for minorities and human rights, and effective political parties.** Trying to force or "rush" democracy on Middle Eastern countries is impractical and counterproductive. The goal should be to help MENA countries develop more pluralistic and representative governments that respect the rights of minorities.

- **Recognize that the key to effective action is local political action, dialogue, education, efforts to use the media, and public diplomacy.** The West and the United States cannot hope to win a struggle for Islam and reform from the outside. It is the efforts of local governments, reformers, educators, and media that will be critical. Encouraging and aiding such efforts is far more important than advancing the image of the United States or Western states or trying to shape local and regional attitudes through Western public diplomacy.

- **Avoid generalizing about Muslims.** Generalizing Islam as a source of violence and discriminating against Muslims in the West can alienate "uncommitted" Muslims.

- **Recognize that demonizing any part of Islam will aid extremists.** The problem of terrorism is not the problem of "puritan" or "Wahabi" Islam, but the attitude of violence and intolerance of politically motivated groups that exploit religious teaching to gain legitimacy in the eyes of their recruits and followers. To defeat these groups, their motivations need to be understood and fought at their roots, e.g., al-Qaeda's goal of ruling the "Arabian Peninsula."

- **Avoid supporting "secularism" against "traditionalism."** The region has seen its share of failed governance systems. Most efforts to secularize have failed and the United States should not be seen as a driving force behind what may be assured failure. Moreover, the word "secularism" translates into "elmaniyah" and is often intermingled with the word "atheism."

- **Do not try to divide and conquer.** The West should stay clear of issues like Sunni-Shi'ite frictions and taking sides with ethic and sectarian groups. It does not serve anyone when they are played against each other. The Iran-Iraq War was a perfect example of how interfering can backfire. The United States should avoid playing any role that could encourage such divisions, particularly given the current environment in Iraq.

- **Understand the difference between liberalism and counterterrorism.** The liberty democratic societies afford people is sometimes the same tool extremists use to spread their hateful ideology. The West must be careful in advocating immediate liberalization of and freedom of speech in the Middle East.

- **Apply a single set of standards to Western and regional counterterrorism.** Do what you preach and preach what you do. The West and specifically the United States should avoid being seen as supporting violation of human rights and abusive security measures in counterterrorism, while advocating human freedom. Violence by states against civilians, be it Russia, Egypt, or Israel, should be equally condemned.

In short, any effective strategy to deal with terrorism and extremism means addressing two key strategic issues that go far beyond the so-called war on terrorism.

One is whether the Arab world can recognize the need for reform and achieve it. The second is whether the West, and particularly the United States, can learn to work quietly with nations for effective reform, rather than seek to impose it noisily, and sometimes violently, on an entire region.

State Support of Terrorism and the Use of Terrorist Proxies

The regional security problems created by independent terrorist movements are further compounded by the state support of terrorism or state use of terrorist proxies. Several states have actively sponsored external terrorist movements or have conducted acts of terrorism outside their own territory. These states have included Iran, Iraq, Libya, and Syria.

Such states may help extremist movements acquire weapons of mass destruction in the future, and the most serious challenge proliferation poses to MENA energy facilities may well prove to be the risk that proliferation interacts with terrorism. At present, this is only a possibility, but terrorist attacks using weapons of mass destruction would present a fundamentally different kind of threat. They would be a far more lethal kind of terrorist threat than the region and the West have yet faced.

Under many conditions, a single act of terrorism can kill thousands of people and/or induce levels of panic and political reaction that governments cannot easily deal with. Under some conditions, the use of weapons of mass destruction can pose an existential threat to the existing social and political structure of a small country—particularly one where much of the population and governing elite is concentrated in a single urban area.

Terrorism and Middle East Energy

Both MENA energy exporters and global energy consumers need a smooth flow of energy exports that must be delivered reliably on a day-by-day basis and be expanded over time to meet global demand. Chapter 1 has shown that the world needs the Middle East and North Africa to both make massive increases in its energy exports and to sustain these at moderate market prices, provide reliable daily deliveries, and avoid any interruptions in supply. Chapter 3 shows that MENA states face immense demographic and economic challenges that require them to earn as much from energy exports as possible, although the era of "oil wealth" has ended and stability can come only from both energy exports earnings combined with a much more diversified pattern of overall economic development.

So far, terrorism and extremism have rarely made direct attacks on energy facilities. This may be because most Islamic extremist movements act largely as national groups or subgroups and see energy export earnings as serving national needs and not just those of the regime or Western interests. There has, however, been a history of minor sabotage in Bahrain and Saudi Arabia, and al-Qaeda has attacked foreign compounds in Saudi Arabia in ways that could have a future impact on the foreign expertise Saudi Arabia still needs for some aspects of its energy production.

Attempts against Saudi oil continue to worry the global energy market and the Saudi leadership. Following a siege and a raid by the Saudi security forces against extremists in Dammam, the Saudi security forces discovered more than 60 hand grenades and pipe bombs, pistols, machine guns, rocket-propelled grenades, two barrelfuls of explosives, and video equipment. The Saudi Minister of Interior, Prince Nayef al-Saud, was quoted saying that the al-Qaeda cell had planned to attack the Saudi oil and gas infrastructure, but Prince Nayef added, "There isn't a place that they could reach that they didn't think about," and insisted that al-Qaeda's ultimate goal was to cripple the global economy.[10]

There have also been occasions in the Algerian civil war when terrorists attacked energy targets and workers in energy facilities. Pipelines and energy facilities were sabotaged during the Iran-Iraq War, although conventional attacks dominated the damage to energy facilities.

There has also been a consistent pattern of systematic terrorist attack and sabotage of Iraq's energy facilities since the U.S. and British occupation of Iraq. Since the start of the Iraq War in March 2003, insurgents have attacked Iraq's oil infrastructure, its pipelines, and its refineries. According to the Institute for the Analysis of Global Security, there have been more than 200 attacks against Iraqi gas and oil installations since June 2003.[11] The insurgents, however, have not succeeded in attacking a major oil field in the north or the south, nor is there evidence to suggest that they have attempted to do so.

It is difficult to generalize from such unique cases, and particularly ones that are still in progress, but the Iraq War has at least shown that pipelines and export facilities are vulnerable and that attacking them can have powerful political and economic effects. These attacks have also shown that their success lies in more than merely destroying the oil infrastructure; it rests on their ability to change how the target country is perceived as a reliable supplier and desirable destination for investment.

Middle Eastern states are also becoming steadily more vulnerable to sabotage and terrorist attacks. Economies of scale lead to the procurement of highly specialized facilities whose equipment involves long lead times for manufacture and repair. Increases in pipeline capacity increase vulnerability, and petrochemical plants often make lucrative targets as do refineries. Attacks on desalination facilities offer extremely lucrative targets that affect the workers in energy facilities. The creation of large, heavily automated gas trains is creating a new target mix in many countries, and electric power is necessary for oil and gas field operations, export facilities, petrochemical production, and civil life. Even comparatively low-value targets like individual oil and gas wells can be attacked in ways that can lead to importer panic or overreaction and force states to deploy large forces to protect entire fields.

There is no way to know how these various forces will play out or how much they will affect energy development and supply from the MENA region. It is clear, however, that they already have significantly increased the risk premium many Western companies see as necessary to invest and do business in the MENA area. They have increased the reluctance to provide foreign investment to an area whose nations have long created legal and economic barriers to such investment, and they have led a number of Western businessmen and technical personnel to leave key MENA energy exporting nations like Saudi Arabia. The Arab world, in turn, is increasingly more reluctant to deal with the United States, and there have been minor boycotts of U.S. companies over U.S. support of Israel. There has been much less reluctance to deal with Europe, but Islamic extremists continue to attack outside investment and Western influences in broad terms, and not just U.S. influence.

There does, therefore, seem to be a growing risk that the forces of extremism and terrorism will present a growing direct threat to energy exports and facilities. The target is extremely tempting. It is one of the few areas where attackers can easily threaten the fiscal stability of the MENA regimes they are seeking to overthrow and have significant leverage against the West. Already, Saudi Arabia has witnessed broader attacks on Western businessmen, and the guerrilla war in Iraq is demonstrating how attacks on Western workers can affect nation building as well as energy supply.

REGIONAL SELF-DEFENSE AND THE ROLE OF OUTSIDE POWER PROJECTORS

The overall military stability of the MENA region is heavily dependent on Western power projection capabilities, and particularly that of the United States. The United States has shown in the Iraq War just how well it can project conventional military power, although it has also shown just how many difficulties it can encounter in dealing with asymmetric warfare and nation building. While the fall of Saddam Hussein has removed a key threat to the region's military stability, threats such as Iran and the wider danger of proliferation of WMD remain.

The present security structure of the MENA region is still dependent on *de facto* alliance between the moderate states in the Middle East and the West and on their access to help from Western power projection capabilities. The United States is the key to such power projection, but this makes it the target of many opposition movements and extremist groups as well.

The high profile of U.S. forces in the Gulf has also interacted with the tensions caused by the Second *Intifadha* and the War in Iraq to cause sufficient political backlash so that the United States is now a major target for terrorists, and it presents growing political problems for U.S. allies, such as Egypt, Jordan, and Saudi Arabia.

European power projection forces cannot substitute for those of the United States. The European members of NATO have never developed the capability for large-scale power projection in the MENA region and are unlikely to do so. The Iraq War has deeply divided the United States and Europe. Europe itself is deeply divided over the attention and role it should play in dealing with the problems in Algeria and North Africa, and the European Union's efforts to create power projection forces have so far been more political than real. Britain and France are the only NATO powers capable of meaningful power projection to the Gulf, but they have minimal strategic lift and only about half the potential pool of forces they had in 1990.

The end result is that the United States continues to play the critical role in defending the major energy exporters in the Gulf, and this is compounded by the fact that the United States is the only power that now can play a major role in securing the global lines of communication to the MENA region and the flow of tankers and other oil exports.

At the same time, major changes are taking place in the regional role of the United States and other Western states in projection power in the Middle East. The U.S. and other Western power projection will be made steadily more complicated by proliferation and the development of more dangerous forms of asymmetric warfare.

The West-MENA Military Relations

The Israeli-Palestinian conflict, the Iraq War, and the problems the United States has had in dealing with allies like Saudi Arabia since 9/11 have made it harder for the United States to maintain a presence and operate in the MENA region. The United States still maintains major forces in the region and many countries depend upon the United States, but U.S. and regional military relations are uneasy at best. Europe talks about power projection capability, but it is not spending enough money on it or developing the systems to ensure broad interoperability with the United States.

The wars in Afghanistan and Iraq have also reinforced the lessons of Lebanon and Somalia that the United States is far from ready to fight asymmetric warfare and highly political conflicts in ways that effectively terminate wars and deal with the issues of peacemaking and nation building. The United States has not shown it can properly characterize and target forces with weapons of mass destruction. Perhaps most important, the United States has never planned to help regional states deal with internal security or save a regime from its own people. U.S. and Western capabilities cannot play a military role in dealing with what may be the most serious threat to MENA stability and energy exports.

Enhancing Cooperation in Counterproliferation and Counterterrorism

The relations between the United States and its allies in the region have started to improve. Although the Arab countries continue to disagree with the U.S. presence in Iraq, there are signs that they are cooperating in counterterrorism and reenergizing their military cooperation. They have also found ways to cooperate on regional issues that are in their mutual interests. For example, Saudi Arabia joined the United States, the United Kingdom, and France and played a major role in pressuring the Syrians to withdraw their troops from Lebanon. These countries have also pressured the Syrians to cooperate with the UN investigation following the bombing that killed former Lebanese Prime Minister Rafik al-Hariri.

In addition, while the United States and its regional allies do not have a common strategy to counter asymmetric warfare and terrorism, they have quietly shared intelligence and cooperated in fighting groups such as al-Qaeda. For example, it was reported that Saudi Arabia provided intelligence to the United Kingdom before the July 7, 2005, bombing in London. The United States and the United Kingdom have also been working closely with Jordanian, Saudi, and Egyptian officials to track and capture major terror suspects.[12]

The situation is more troublesome in terms of proliferation. The United States, its Western allies, and its allies in the region do not yet have a clear counterproliferation option or surplus funds to pay for such an option. As mentioned earlier, the MENA countries have consistently called for a nuclear-free Middle East and have tried to link Iran's nuclear ambition with Israel's nuclear arsenal. For example, the GCC Secretary General asserted that the GCC is being "sandwiched" between two nuclear powers, Israel and Iran. He went on to say, "I call on NATO to exercise direct pressure to eliminate WMDs (weapons of mass destruction) from our region, without exception."[13] Asymmetric warfare and proliferation do remain the major two threats to stability in the MENA region. They also have the most direct threat to energy production and export routes in the Gulf.

ENERGY VULNERABILITY AND MARITIME CHOKEPOINTS

Given the interdependence of the global energy market, perceptions of instability are as important as realities. To reassure markets, producers have to build confidence not only in their capability to prevent attacks, but also in their ability to contain the damage of unexpected violence through building redundancy in production and export systems. In addition, as discussed in detail earlier, the overall geopolitical risks surrounding MENA oil can take many forms. The security of oil facilities is most obvious and has the most immediate damage to the global economy and stability of sovereign governments in the region. In 2005, the International Energy Agency (IEA) summarized its apprehensions about energy security as follows:[14]

> [S]erious concerns about energy security emerge from the market trends...The world's vulnerability to supply disruptions will increase as international trade expands. Climate destabilizing carbon-dioxide emissions will continue to rise, calling into question the sustainability of the current energy system. Huge amounts of new energy infrastructure will need to be financed. And many of the world's poorest people will still be deprived of modern energy services. These challenges call for urgent and decisive action by governments around the world.

In recent history, energy supply disruptions that have happened in the MENA region have been the result of attacks, political motivation, and lack of management. Supply disruptions are scarcely unique to the MENA region. Other exporters in Africa, Central Asia, and Latin America already have more serious internal security problems. According to Ian McCredieo, the head of Shell's Global Security Services, 14 oil-producing regions have local security forces that are "largely ineffective." Oil companies have experienced kidnappings from rebel and guerrilla groups in Africa and Latin America, threat of terrorist attacks in the Middle East, danger of piracy of oil tankers in the Malacca Straits, and expropriation and "industrial espionage" in Russia.[15]

Most foreseeable assaults against energy infrastructure, however, are likely to be quickly confined, and any resulting damage is likely to be repaired relatively quickly. Energy security, however, will be a continuing problem particularly if global energy demand does actually rise by more than 50 percent by 2025. The security of energy exports will play a steadily more vital role in the world's economy.

The Middle East makes a good case in point. In the last two decades, the Middle East and Africa experienced the largest increase in oil production of any other region. According to British Petroleum, in 1983, Africa produced 4.9 million barrels per day (MMBD) and the Middle East produced 11.8 MMBD, but by 2004, these numbers reached 9.3 MMBD and 24.6 MMBD, respectively. Production levels of other regions remained at approximately the same levels they were in the 1980s.[16]

History of Oil Interruption and Embargoes

Regional stability is multilayered. Social and political stability is as important as effective security. After all, the recent price hike to $70 per barrel was caused by a surge in demand compounded by labor strikes in Venezuela, hurricanes in the Gulf of Mexico, the Iraq War, and the ongoing disruptions of Angolan and Nigerian oil compounded by the surge in oil demand.

The threats of oil and gas supply disruptions show that the risk of a serious interruption in Middle Eastern energy exports, and particularly Gulf exports, cannot be ignored. As Figure 2.16 shows, there has been a long history of oil interruptions since 1951. Virtually all of these interruptions have taken place in the MENA region, and some have been serious.

The oil embargo of 1973–1974, for example, triggered a massive rise in oil prices that reshaped the energy costs of the global economy and made dependence on energy imports a major strategic issue for the first time. The fall of the Shah in 1979, and the Iran-Iraq War that followed, created a global panic in the oil market and again dramatized strategic dependence on the Gulf. The period of 1979–1980 also marked the height of MENA energy export revenues in constant dollars. Iraq's invasion of Kuwait and the Gulf War of 1990–1991 marked another major rise in oil prices, although its affects were much less serious than the interruptions of 1973–1974 and 1979–1980.

Security of Energy Infrastructure

If one observes the incidents outlined above, there are no incidents of attacks against oil fields or energy infrastructure per se. The Iraq War might be the only incident where there was physical damage to the oil infrastructure, and it is often pipelines and other export facilities. This is mainly due to the fact that while oil fields are large-area targets, they have many redundant facilities. While fires can be set in many areas of a working field, including at oil wells, fires do not produce critical or

Figure 2.16   Global Oil Supply Disruptions: 1951–2004

Date of Oil Supply Disruption	Duration of Supply Disruption (in Months)	Average *Gross* Supply Shortfall (in MMBD)	Reason for Oil Supply Disruption
3/51–10/54	44	0.7	Iranian oil fields nationalized May 1, 1951, following months of unrest and strikes in Abadan area.
11/56–3/57	4	2.0	Suez War
12/66–3/67	3	0.7	Syrian Transit Fee Dispute
6/67–8/67	2	2.0	Six Day War
5/70–1/71	9	1.3	Libyan price controversy; damage to Tapline
4/71–8/71	5	0.6	Algerian-French nationalization struggle
3/73–5/73	2	0.5	Unrest in Lebanon; damage to transit facilities
10/73–3/74	6	2.6	October Arab-Israeli War; Arab oil embargo
4/76–5/76	2	0.3	Civil war in Lebanon; disruption to Iraqi exports
5/77	1	0.7	Damage to Saudi oil field
11/78–4/79	6	3.5	Iranian revolution
10/80–12/80	3	3.3	Outbreak of Iran-Iraq War
12/02–2/03	3	2.1	Venezuela strikes and unrest.
3/03–8/03	6	0.3	Nigeria unrest.
3/03–9/04	19	1.0	Iraq War and continued unrest.

Source: EIA, "Global Oil Supply Disruption Since 1951," available at http://www.eia.doe.gov/emeu/security/distable.html

lasting damage. Unless wells are attacked with explosives deep enough in the wellhead to result in permanent damage to the well, most facilities can be rapidly repaired.

There are, however, larger items of equipment and central facilities whose damage would do far more to interrupt production, and many of which require months of manufacturing time to replace. Such facilities include central pumping facilities, gas-oil separators, related power plants, water injection facilities, and desalination plants. Vulnerability also increases sharply if key targets in a field are attacked as a system, rather than as individual elements, and if expert assistance is available to saboteurs or attackers.

While a nation like Iran can pose a conventional and asymmetric military threat to oil facilities in the Gulf, the more dangerous threat is that of asymmetric attacks by extremists groups on oil facilities in the Gulf, where 65 percent of the world's "proven" reserves exist. There is no attack-proof security system. It may take only one attack on Ghawar (onshore) or Safaniyah (offshore) to seriously destabilize the world energy market.

Security of Export Routes

It may also take an attack on just one export route to throw the global oil market into a spiral. Oil and gas move in many different ways. Oil and gas pipelines connect

North Africa to Europe and may eventually connect Middle Eastern states to South Asia. The Energy Information Administration (EIA) estimates that 90 percent of oil exported from the Gulf is shipped via tankers through the Strait of Hormuz, and the remaining 10 percent through the pipelines.

In addition, most of the other MENA oil and gas are directly or eventually moved by sea. For example, oil was exported out of maritime ports in the Gulf and Indian Ocean and through the Saudi East-West pipeline to the port of Yanbu on the Red Sea, via pipeline from Iraq's Kirkuk oil region to the Turkish port of Ceyhan, and by pipeline via Syria. Only comparatively small amounts to their ultimate destination moved by land, largely by truck to destinations like the Kurdish areas of northern Iraq, Turkey, Jordan, and Iran.[17]

It is easy to focus on the security of oil and gas fields, energy facilities, and pipelines in the MENA area and to forget that most energy exports ultimately move by sea. Moreover, Chapter 1 has indicated that Gulf exports alone will require something like 2.5 times the tanker traffic by 2025 that exists today, as well as vastly expanded ports and loading facilities. Much of this increase in tanker traffic will go to Asia and through the Indian Ocean and the Pacific, but a substantial portion will go to Europe and the United States.

The flow of oil exports can be attacked at any point during a tanker voyage. However, there are several key maritime chokepoints that could have a critical impact on the flow of oil in the Middle East. The EIA summarized the three chokepoints in the MENA region as follows:[18]

- **Strait of Hormuz:** The Strait consists of two-mile wide channels for inbound and outbound tanker traffic, as well as a two-mile wide buffer zone. The oil that flows through the Strait of Hormuz accounts for roughly two-fifths of all world traded oil, and closure of the Strait of Hormuz would require the use of longer alternate routes (if available) at increased transportation costs. Such routes include the approximately 5-MMBD-capacity East-West Pipeline across Saudi Arabia to the port of Yanbu, and the Abqaiq-Yanbu natural gas liquids line across Saudi Arabia to the Red Sea. The 15.0–15.5 MMBD or so of oil which transit the Strait of Hormuz go both eastward to Asia (especially Japan, China, and India) and westward (via the Suez Canal, the Sumed Pipeline, and around the Cape of Good Hope in South Africa) to Western Europe and the United States.

- **Bab al-Mandab:** Oil heading westward by tanker from the Gulf toward the Suez Canal or Sumed Pipeline must pass through the Bab al-Mandab. Located between Djibouti and Eritrea in Africa, and Yemen on the Arabian Peninsula, the Bab al-Mandab connects the Red Sea with the Gulf of Aden and the Arabian Sea. Any closure of the Bab al-Mandab could keep tankers from reaching the Suez Canal/Sumed Pipeline complex, diverting them around the southern tip of Africa. This would add greatly to transit time and cost, and effectively tie up spare tanker capacity. In December 1995, Yemen fought a brief battle with Eritrea over Greater Hanish Island, located just north of the Bab al-Mandab. The Bab al-Mandab could be bypassed by utilizing the East-West oil pipeline. However, southbound oil traffic would still be blocked. In addition, closure of the Bab al-Mandab would effectively block nonoil shipping from using the Suez Canal, except for limited trade within the Red Sea region.

- **Suez/Sumed Complex:** After passing through the Bab al-Mandab, oil en route from the Gulf to Europe must pass through either the Suez Canal or the Sumed Pipeline complex in Egypt. Both of these routes connect the Red Sea and the Gulf of Suez with the Mediterranean Sea. Any closure of the Suez Canal and/or the Sumed Pipeline would divert tankers around the southern tip of Africa (the Cape of Good Hope), adding greatly to transit time and effectively tying up tanker capacity.

These chokepoints, like the Strait of Hormuz, remain critical to energy security. At the same time, the proliferation of long-range naval strike aircraft, antiship missiles, smart mines, submarines, and guided missile ships is extending the range at which energy exports can be threatened.

These changes in military technology and in the flow of Gulf exports are changing the definition of "chokepoint." One key example is the acquisition of long-range missiles and weapons of mass destruction by nations like Egypt, India, Iran, Iraq, Israel, Libya, and Syria. Another is Iran's development of bases on islands near the Strait of Hormuz and its long-standing dispute with the United Arab Emirates over the three islands near the Strait. In addition, Iran's acquisition of advanced antiship missiles, submarines, long-range strike aircraft, and missile patrol boats poses a threat to oil tankers. The same weapons and technologies allow any nation along the shipping lanes to Asia to create new "chokepoints" at ranges up to several hundred kilometers.

THE PROBLEM OF GUESSING AT FUTURE SCENARIOS

There is no consensus as to what kind of interruptions might take place in the flow of MENA oil exports, in part because there is no current set of contingencies or threats that appears probable enough to merit detailed planning. The preceding analysis has shown that such interruptions could have a wide range of causes and take a wide range of forms.

Analyzing future interruptions, therefore, requires taking into account much of the uncertainty surrounding the energy market from a geopolitical angle. The links between geopolitical risks and production uncertainties may seem tenuous to laymen but are nonetheless central to stability in the energy markets and the energy-producing states. The following are key areas of uncertainty that make guessing at future scenarios hard:

- **Stability of oil- and gas-exporting nations:** The stability of oil- and gas-producing nations is of paramount importance to the world oil and gas markets. The strikes in Venezuela, the War in Iraq, and the ongoing disruptions of Angolan and Nigerian oil are examples of what could happen if this happens in other countries such as Saudi Arabia and Iran.

- **Terrorism in the Gulf and oil facilities securities:** While the threat from Iran's conventional military may be real, the more dangerous threat may be that of extremists groups' asymmetric attacks on oil facilities. The Gulf contains over 65 percent of the world's proven reserves, and, as mentioned earlier, there is no attack-proof security system.

- **Proliferation of WMD:** The success in stopping Abdul Qadeer Khan does not mean the end of a nuclear black market. It remains a real threat to the entire world, especially the Gulf, of a nuclear weapon falling in "the wrong hands," such as al-Qaeda.

- **Embargoes:** The prospects of a new embargo by oil producers seems unlikely today. In the 1973–1974 embargo, the world market increased oil production to compensate for the cuts in Middle Eastern exports, though world markets were not as capable of tracking what was happening or effectively identifying and distributing the oil available. As a result, the seriousness of the crisis was exacerbated by the world's inability to deal with it. But the world has become better at handling interruptions when they occur. Neither the "tanker war" between Iran and Britain and the United States in 1987–1998 nor the Gulf War in 1990–1991 led to the same level of panic, price rises, and hoarding.

- **Sanctions:** The probability that the international community or a group of countries will impose sanctions on oil exporters remains credible. Iran's nuclear weapons programs render it a possible target of sanctions, as do the actions of the Sudanese government in Darfur. Although there is no international consensus as of yet to levy sanctions on these two countries, or that sanctions would cover oil exports, there is a certain ambiguity that contributes to the overall uncertainty of the world oil market.

- **Ethnic conflicts and sectarian strife:** Disagreements over the control of oil revenues by ethnic tribal, religious, and other factions can destabilize countries and disrupt the flow of petroleum and gas. Currently, the ongoing conflict in the Niger Delta and the War in Iraq provide two examples of how devastating such crises are.

- **Natural disasters:** Natural incidents in production, export, or refining areas can be damaging to the energy market. Hurricanes in the Gulf of Mexico have caused supply and distribution disruption in the United States and have added large premiums to the price of a barrel of oil. Hurricanes Katrina and Rita, which hit the United States during August and September 2005, shut down most of the refineries in the U.S. Gulf of Mexico and forced the United States to release some of its strategic petroleum reserves. They also had a major impact on domestic gas production and prices and the need for imports.

- **Security problems and accidents:** The world can absorb the problems created by most forms of local conflict and internal security problems when there is significant surplus capacity of energy exports, and prices start from a relatively low base. Behavior changes drastically, however, when supply is very limited and prices are already high. Even potential threats to petroleum production, exports, and distribution can radically alter prices and market behavior. Actual attacks, or major industrial accidents, can have a much more serious impact. The loss of a major supplier, or a sustained major reduction in regional exports, potentially can have unpredictable price and supply impacts that affect the entire global economy.

The most likely interruptions seem to be ones that are short or limited in scope, and many could be dealt with by production increases by other countries. The impact of the loss of Iraq and Kuwait oil production during the Gulf War in 1990–1991, for example, was limited by increases in Saudi and other production. Increases in production by other exporters largely compensated for a political crisis that led to Venezuelan production cuts in 2002; the same was true when Iraq ceased to export during the Iraq War of 2003 and during the interruption due to Hurricanes

Katrina and Rita in the Gulf Coast of the United States. While each interruption did produce price rises, and had some impact on global economic growth, the impact was too limited to have a major impact on the global economy.

The steady increases in world demand for MENA energy exports projected through 2030 do mean, however, that the global economy will become steadily more vulnerable to major interruptions. The Gulf alone is projected to more than double its flow of exports during 2000–2025. Five Gulf countries—Saudi Arabia, the United Arab Emirates, Iraq, Iran, and Kuwait—will become steadily more important producers and exporters. At the same time, the gap between the normal production of oil and total production capacity is expected to shrink steadily, leaving less and less surplus production capacity that the MENA region and other nations can bring on line in the event of a major interruption in exports from one or more Gulf States.

It is all too clear how critical the Gulf is to world production capacity, and it is clear just how much the world will come to depend on a steady and stable increase in Saudi production and exports. In fact, Russia is the only major exporter outside the MENA area where a major interruption could have a critical impact on the global economy, particularly because it is both a major oil and gas exporter.

At the same time, Chapter 3 shows that MENA states already desperately need the cash flow from their energy exports, and this need will grow over time. The demographic and economic pressures on the region are so severe that no regime in an exporting country can ignore the consequences of an embargo, and the factions in any civil fighting must consider the popular reaction to an attack on such facilities.

Wars tend to do limited damage of limited duration, not produce catastrophic interruptions, and any belligerents in the region must consider the fact that U.S. and outside intervention would almost certainly occur in the event of a major interruption, just as it did during the Iraq-Iraq War when the United States "reflagged" tankers and defended Gulf shipping against Iran.

In short, there are so many different real-world ways in which an interruption could develop and play out in terms of its politics, the use of military force, the actual level of cuts in exports, in the duration of such cuts, in the way other exporters can compensate, and in terms of the global economic climate and level of demand that such scenario analysis can be little more than a matter of informed guesswork.

The Economic Impact of Energy Interruptions

There is no reliable way to measure the economic impact any given interruption in MENA energy exports would have, and such an impact would vary sharply according to scenario and duration. Under many conditions, an interruption might be limited enough—and be of such short duration—that it would actually have less impact than the normal fluctuations in oil prices that grow out of market conditions.

This price volatility may at first seem so massive that it must have dramatic effects, but it is not atypical of the impact of "normal" market conditions in energy exports. In fact, any analysis of the past ups and downs in oil prices caused by market forces shows that it would take a very serious interruption to have more serious effects. It is important to note, however, that the shifts in prices during 1997–2001 occurred over an extended period of time and avoided panic buying. It is almost impossible, however, to estimate the psychological impact of sudden interruptions on the market, and it is clear that major market-driven changes in energy prices—which would be similar to the impact of a major oil interruption—had a real but moderate impact on the U.S. economy.

The EIA discussed historical impacts of high energy prices on the U.S. economy in the following analysis:[19]

- Viewed from a long-term perspective, inflation, measured by the rate of change in the consumer price index (CPI), tracks movements in the world oil price. Not only do oil and other energy prices constitute a portion of the actual CPI, but downstream impacts on other commodity prices will have a lagged effect on the CPI inflation.

- Looking from the 1970s forward, there are observable, and dramatic changes in GDP growth as the world oil price has undergone dramatic change. The price shocks of 1973–1974, the late 1970s/early 1980s, and the early 1990s were all followed by recessions, which have then been followed by a rebound in economic growth. The pressure of energy prices on aggregate prices in the economy created adjustment problems for the economy as a whole.

- In the past year, forecasters have acknowledged that higher energy prices can become a drag on the overall economy. Initially, overall CPI inflation was still very low, principally because inflation in commodities other than energy and agriculture was extremely low. However, the sustained high level of oil prices has begun to affect core inflation (minus energy and food) through continued pressure on prices of other commodities, in the United States and worldwide.

The economic impact of an interruption could, however, be reduced if the United States made timely and effective use of its Strategic Petroleum Reserves (SPR) and the interruption was not serious enough to trigger the sharing of available oil imports called for under the security agreements the United States and its allies signed in creating the International Energy Agency. The SPR currently has a storage capability of 727 million barrels.

The SPR played a key role in stabilizing the U.S. energy market and preventing major shocks to the U.S. economy following the disruption to refineries during Hurricanes Katrina and Rita, which pushed the price of oil to above $70/barrel. The U.S. Department of Energy and the IEA coordinated and authorized the sale of 30 million barrels of crude oil from the SPR.

The long-term economic impact of Katrina and Rita, however, might not yet be fully known. Agencies such as the EIA and the IEA have attempted to model these shocks, and their forecasts tend to take into account such idiosyncrasies. They are by "nature" hard to detect, hard to measure, and no one fully knows their level of persistence. This leaves many modelers with the creating "best guesstimates" or ranges where they think the impact can influence the energy markets and the global economy.

The EIA "Rules of Thumb" for Calculating the Impact of Energy Interruptions

The EIA, for example, has established some rough rules of thumb for estimating the impact of major interruptions on the macroeconomic level. They also provide a range of measured uncertainty or a best benchmark in the case of supply disruption. These rules of thumb are updated to take into account recent developments. Given the recent surge in oil price, the EIA has divided its rules of thumb into the following and raised its high-price case to include $50/barrel:[20]

- **If the price is around $20–$30/barrel:** the rule of thumb is a price increase of $3–$5 per barrel per million barrels per day of net supply disruption.
- **If the price is around $40/barrel:** the rule of thumb is a price increase of $4–$6 per barrel per million barrels per day of net supply disruption.
- **If the price is around $50/barrel:** the rule of thumb is a price increase of $5–$7 per barrel per million barrels per day of net supply disruption.

The EIA argues that the reason for these rules of thumb is to maintain similar percentage price increases necessary to balance supply and demand in the market in case of supply disruptions. The EIA also sets rules of thumb for the impact of rise in the prices of oil on the U.S. GDP. The EIA asserts that a 10-percent increase in the price of oil could lower the real U.S. GDP growth rate by between 0.05 and 0.1 percentage points, and it argues that the impact in the first year will be closer to 0.05 percent where after the first year, the impact is closer to 0.1 percent.

These provide important benchmarks, and as the EIA noted, these rules of thumb are subject to important qualifications:[21]

First, estimate how much oil was being produced in the disrupted countries that is no longer available. EIA defines this as the gross disruption size. Then, to better estimate the price and economic impacts of a given oil supply disruption, subtract from the gross disruption size the amount of extra oil that unaffected countries are likely to produce to help offset the loss of oil to the market. Oil prices are likely to increase immediately with the outbreak of the disruption. Over time, however, the higher price increases the incentive for other producing countries, where possible, to increase their oil production. After subtracting extra production from the gross disruption size, the result is what EIA calls the net disruption size. Using the net disruption size in the rules-of-thumb listed above provides a rough estimate regarding potential impacts of a given disruption. Other factors, such as the level of commercial oil inventories and the use of such stockpiles as the Strategic Petroleum Reserve, will also influence the price and macroeconomic impacts of any given oil supply disruption.

The grim truth is that the nature and economic impact of any given energy interruption is likely to be known only as it develops. It is always possible to speculate using both the scenarios discussed earlier and the EIA rules of thumb, but the chances of their coinciding with reality are negligible. This does not, however, make the risks less real. The fact that no one can predict the impact of regional conflict and instability 30 years into the future, or even the nature and the outcome of the most likely cases, has never prevented them from occurring.

Political, Economic, and Demographic Dynamics in the MENA Countries

Like most parts of the world, the countries in the Middle East and North African (MENA) region suffer from very real internal instabilities. These include internal political instability, foreign conventional military threats, the proliferation of weapons of mass destruction (WMD) threats, asymmetric threats, violent Islamist extremism, and the threat of terrorism. They also include civil threats like overdependence on energy export revenues and economic stagnation, religious and cultural alienation, youth and demographic pressures, and ethnic and sectarian divisions.

On the political side, countries need to strengthen the rule of law, both in human rights and the basic protection of property, civil proceedings, and economic activities. They lack a stable popular base and the representative governments that serve the nation. From the economic dimension, most of the MENA countries—the oil-exporting nations—are highly dependent on oil and lack real prospects for realistic economic diversification policies. They also lack vibrant private sectors, which is largely the result of strong state control over the economic activity and that has contributed greatly to the high rates of unemployment across the region.

Many of these problems are self-inflicted wounds that have been in the making for decades. This is particularly true of the slow pace of political change and reform, the economic problems in the region, and the impact of rapid population growth. While it is true that outside players have affected the strategic landscape of the MENA region, most of the causes of instability stem from the actions of the nations in the region. Far too often, regional leaders and intellectuals have failed to come to grips with the true causes of instability and blamed outside powers for their own actions (or inaction).

There is no way to predict when, if ever, given sources of instability will have a major or lasting impact on energy production and exports or in maintaining and expanding energy production capacity. It is easy to postulate worst-case scenarios, and it is just as easy to argue that nations have coped in the past and will cope in the future. The region's recent history, however, is one of very different nations with very different behavior. It is also one of sudden crises that few anticipated in the form in which they actually occurred and whose impact on energy supply then proved to be just as unpredictable as the events that unfolded.

If there is a lesson from such history, it is that risk needs to be constantly reevaluated on a national level. At the same time, there are forces at work that do operate on a regional level. For all the differences between MENA nations, there are common political, economic, and demographic pressures that may well have a serious impact on either the flow of energy exports or efforts to maintain and increase export capacity.

POLITICAL UNCERTAINTY

Most Middle Eastern and North African political structures remain delicate, regardless of the formal structure of the government involved. Many states are making progress, but no state in the region has as yet managed to create a political culture that provides effective pluralism to the many ethnic, religious, and tribal groups while implementing meaningful enforcement of the rule of law and respect for human rights. All of the competing ideologies of the postcolonial era have so far failed: Pan-Arabism, socialism, capitalism, Marxism, and statism have not provided lasting political cohesion, given development adequate momentum, or met social needs.

As for Islam, the fact that much of the population of the region has turned back to more traditional social structures and religion is scarcely surprising. It also remains a fact that many of the Islamist movements, groups, and political parties have taken the role of providing many social, health, and educational services where some governments have failed to deliver. Many experts believe that this is one of the reasons that groups such as the Muslim Brotherhood in Egypt and Syria, Hamas in Palestine, as well as other Islamist groups in North Africa are seen by the public as more effective providers of services. Politics in the Gulf, where oil wealth is high, have not been as influenced by the ability of Islamist groups to provide educational, welfare, health, and other services. There is, however, a very real ideological struggle for power.

More generally, the tensions caused by Islamist extremism have been compounded by the fact that the region has failed to build meaningful political institutions. Even in most countries that claim to hold elections, the ruler or a one-party system dominates. Political parties and elections rarely allow open competition and campaigning. That is not to say that politics does not take indigenous forms through tribal, religious, or other civil society elements, but it often lacks the institutions necessary for long-term stability. Real political parties, free speech, the rule of law, and the basic security that should be part of human rights are often limited or missing.

That being said, the region is scarcely without hope. Leaders have emerged in some MENA countries that are pressing for serious reform, such as building viable political parties and civil society institutions. Far too often, however, MENA societies are so static that they may be moving toward revolution or civil war where they should be moving toward evolutionary political and economic reform. The tragedy of the Middle East is that so many opportunities are being wasted and the region is steadily falling behind the cutting edge of political, economic, and social development in areas like Asia and Latin America.

Uncertain Political Succession

The impact of the region's political problems must also be kept in historical perspective. Middle East and North African regimes have been remarkably politically stable. Despite opposition groups, coup attempts, and a number of low-intensity conflicts, regimes in the MENA region have ruled uninterrupted for decades. Some of the regional leaders ruled uninterrupted for many decades, such as King Hussein of Jordan (1953–1999), Sheikh Issa al-Khalifah of Bahrain (1961–1999), King Hassan of Morocco (1961–1999), Hafez al-Assad of Syria (1971–2000), Sheikh Zayed Al-Nuhayan of the United Arab Emirates (1971–2004), Sheikh Maktoum bin Rashid Al Maktoum of Dubai (1990–2006), King Fahad al-Saud of Saudi Arabia (1982–2005), Sheikh Jaber al-Sabah of Kuwait (1977–2005), Muamar Qadhafi of Libya (1969–Present), and Sultan Qaboos of Oman (1970–Present).

Many other countries, such as Saudi Arabia, have gone through a succession process that is smooth without violence or dispute among the ruling elites. Peaceful successions have taken place in Iran, Jordan, Bahrain, Morocco, Syria, the United Arab Emirates, and Saudi Arabia. The successful overthrows of a ruler since the fall of the Shah of Iran have also taken place without any meaningful economic impact: the current Qatari emir, Hamad Khalifah al-Thani, overthrew his father, Khalifah al-Thani, in 1995; and the military overthrew President Maaoya Sid'Ahmed Taya of Mauritania in August 2005 in another bloodless coup. For all of its problems, the region has recently been more stable than it was at the time of Pan-Arabism and Gamal Abd al-Nasser, the Iranian Revolution, the Gulf War, and the Lebanese Civil War.

One exception to this rule has been the recent quarrel between Sheikh Sabah al-Ahmad al-Sabah, Kuwait's Prime Minister, with Crown Prince Sheikh Saad al-Abdullah about who should succeed Sheikh Jaber al-Sabah as Kuwait's ruler. While succession was usually a family affair, the Parliament was actively involved in the discussions about who should succeed Sheikh Jaber; in the end, Sheikh Sabah al-Ahmad al-Sabah was elected emir given that Sheikh Saad al-Abdullah was deemed too sick to hold office. This dispute underlined the uncertainties involved in the political succession in many of these countries, compounded in Kuwait's case by vibrant parliamentary politics; but this story also underscored the resilience of the ruling families in smoothening the transition of power.

Key Area of Political Uncertainty in the MENA Countries

It is far too early, however, to draw any conclusions about future successions, regime stability, or the political nature of the MENA states. Even if one disregards Islamist extremism, there is no guarantee of stability or instability for a future that extends out to the year 2020. Each country has its own set of political uncertainties, which may determine the future of the individual countries. In addition, some of these trends may influence the region as a whole. The key domestic political issues affecting oil- and gas-exporting states include the following:

- **Sunni-Shi'ite divisions in Bahrain and its tenuous political stability:** Bahrain is divided between the Sunni ruling elite and the 90 percent of the population who are Shi'ites. These tensions have grown in recent years in spite of efforts by the current king, King Hamad, to introduce reform. There have also been divisions within the al-Khalifah royal family over how to govern and deal with the growing gap between Shi'ites and Sunnis. The divisions have rarely been violent, but ongoing uncertainties raised by possible Iranian involvement and the rise of Iraq's Shi'ites may add to the pressure from Bahraini Shi'ites on Bahrain's political establishment.

- **The changing nature of politics in Kuwait and its uncertain succession future:** Kuwait faces an uncertain succession process. The main issue is the fact that historically, the succession has switched from two different branches of the al-Sabah family. After Sheikh Sabah al-Ahmad al-Sabah became Kuwait's new Emir in January 2006, he named his brother, Sheikh Nawaf al-Ahmed al-Sabah as Crown Prince, and his nephew, Sheikh Nasser Muhammad al-Ahmed al-Sabah, as Prime Minister. Both are from the al-Jaber branch of the al-Sabah family, signaling a break with the tradition of alternating rulers between the two branches of the al-Sabah family. The evolution of this family struggle is compounded by deep political, ethnic, and ideological divisions within the National Assembly; the Assembly is increasingly polarized between Islamists and narrow, special-interest, service politics.

- **The moderates vs. the traditionalists within Iran's ruling elites:** There is an uncertain balance of power between the "moderates" of former Presidents Khatami and Rafsanjani and the "traditionalists" of the Ayatollah Khamenei in Iran. Many experts regard Iran's new President, Mahmoud Ahmadinejad, as a hard-liner. He has strongly defended Iran's right to a full nuclear fuel cycle and has challenged Israel's right to exist. He has also been opposed by the Parliament on key appointments, as three nominees for Minister of Petroleum were rejected before Kazem Vaziri-Hamaneh was confirmed, more than four months after Ahmadinejad came to power in August. This may lead to further divisions across ideological lines, and given the international scrutiny of Iran, these divisions may be exacerbated over Iran's nuclear program. Furthermore, the impact of oil revenues on political stability will do much to determine the ability of Iran's leaders to deal with low economic growth rates, high unemployment, and high population growth.

If Iran continues to proliferate and actively seek nuclear weapons, it may also provoke a major military confrontation with the United States or Israel. At this point in time, it is the only MENA energy exporter with a significant probability of becoming involved in a new war, one where military action against Iran might provoke it to retaliate by action in Iraq, the Gulf, or against Israel.

- **Iraq's political future:** The security situation in Iraq is as uncertain as its political structure. While the Iraqi people have approved a constitution and elected a fully sovereign government, it is uncertain that Iraq can create a stable postautocratic regime. Iraq is a state with a major ongoing insurgency, massive internal economic problems, high population growth, and deep divisions between Sunnis, Shi'ites, and Kurds. These problems are compounded by the fact that the Shi'ite and Kurdish populations are growing much more quickly than the Sunni population, and the sectarian divisions are regional, where the majority of the Sunni population lives in the central part of the country compared to the north and the south of the country which are rich natural resources. Iraq may well be able to create a more national and inclusive government and develop strong enough Iraqi forces to defeat the insurgency. It could, however, become involved in a far more serious civil war.

- **The uncertainty of succession in Oman:** The succession process in Oman is not clear. Sultan Qaboos is in his early late 60s and there is no obvious successor. He has ruled since he took over from his father in 1970. He does not have any sons, but has a number of cousins. The Basic Law of 1996 is equally unclear, but he has promised to name his successor in a sealed letter. This uncertainty is compounded by other internal political problems and rivalries and by the fear that the Sultan is increasingly isolated from the people and is not getting effective technocrat advice.

- **Residual divisions within Qatar's royal family and Qatar's tension with its neighbors:** Qatar has a tendency to provoke "generational" quarrels with Bahrain, Egypt, Jordan, Saudi Arabia, and the United Arab Emirates. The launching of al-Jazeerah network has also upset many neighboring countries and is seen by them as a way to turn their populations against their regimes. There is wide resentment within the ruling family for the 1995 coup, and the Emir remains unpopular among the Qatari elites and the religious establishment for his stance on the Palestinian issue and for allowing the air campaign of the Iraq War to be waged from the Doha air base. Qatar's wealth has bought it a lot of time to deal with its growing debt, large population growth, and the divisions among its ruling elites. The future remains uncertain as to how the ruling family will manage these difficulties.

- **Saudi Arabia's succession to the next generation of royals:** The succession between King Fahd and King Abdullah was smooth. The same is likely to apply to the succession from King Abdullah to Crown Prince Sultan. Many experts question how the succession process will work between this generation of royals and the next one. Some experts, however, believe that the next generation of royals is competent and, in fact, has been playing major roles in leading ministries of the Kingdom. In addition, many of young royals and the new generation of technocrats have been promoted to higher positions in the country. On the security side, Saudi Arabia has faced problems with its al-Qaeda and other extremists groups, and it continues to struggle with high unemployment rates coupled with a "youth explosion" due to high population growth rates.

- **Political rivalries in the United Arab Emirates:** Divisions within the UAE royal families remains unclear, especially after the death of Sheikh Zayed in November 2004. The current leader, Khalifah the son of Zayed, does not have popular support. The two emerging leaders are the Emir of Dubai and the Crown Prince of Abu Dhabi. They hold important positions: the former is Vice-President and Prime Minister of the United Arab Emirates and the latter is Joint Chief of Staff. Generally, the next generations of leaders do not agree on major policy issues, including the inflow of foreign labor into the United Arab Emirates, U.S.–UAE relations, the laissez-faire social and economic policies of Dubai, and the Emirates' relationship with its Gulf neighbors.

- **Syria's political vacuum:** The ability of Bashar al-Assad to govern in Syria is decreasing with time. While the old guards of his father's era have helped him in running the Syrian affairs so far, the alleged involvement of Syrian security services in the Hariri assassination and the regime's inability to reform the economy and deal with high population growth are exerting a lot of pressure on the Syrian government. Syria's official involvement in Lebanon came to an end after pressure from the UN and the international community. Furthermore, the divisions among the Alawites, Sunnis, Shi'ites, secularists, and Kurds might be enhanced by the sectarian divisions in Iraq and may complicate the regime's ability to control them. All of this will depend on knowing who is actually running the show, which has often been unclear.

- **Egypt's shifting political landscape:** Egypt allowed its first multiparty election in 2005. President Hossni Mubarak won an expected victory as the President of Egypt. His age— and the lack of any formally designated Vice President in Egypt—do, however, continue to create major uncertainties about the future political stability in the country. The legislative election also exposed far more tension and opposition than many predicted. The Muslim Brotherhood, while outlawed as a formal political party, proved to be the best organized *de facto* party. It also proved to be the best equipped to lead the municipal as well as parliamentary elections. Many are worried about its growing importance in Egypt. These political uncertainties have been compounded by reported rumors about the ambitions of Mubarak's son to become the next President of Egypt.

- **Tunisia's uncertain path toward stability:** Tunisia has made strides in reforming its economy and at least in starting the building of a middle class. The overall prospects for stability are good, but President Zaine El-Abidine Ben Ali's health remains the major question. In addition, he has failed to build a stable base of pluralism for his regime, and he has alienated many of the moderate Islamist oriented groups with his overt secularist drive, which some experts believe can drive the moderates to become hard-liners over time. These issues raise serious questions about the future of Tunisia.

- **Libya's changing social, political, and strategic cohesion:** Libya is no longer subject to sanctions and has improved its relations with the United States. However, Qadhafi faces problems. His relationships with other governments have not been as solid following the revelation that he was involved in an alleged plot to assassinate then Crown Prince Abdullah of Saudi Arabia. From a political point of view, his son, Saif, is becoming more visible and is seen by many as his successor. His health has not been a problem, but he is old and there is not a formal succession procedure. In addition, he is facing a growing challenge from Islamists who are unhappy about his autocratic rule and especially his "secular-nationalist" tendencies.

- **Algeria's fragile truce:** Religious extremism and secular corruption and authoritarianism have proven to be an explosive mix in Algeria. The civil war, however, has given way to a period of relative calm. President Abdulaziz Boutafliqah was elected in 1999 and was reelected in 2004—in what many international observers saw as a fair election. Many experts believe that security and political stability have improved in the last six years. Boutagfliqah's amnesty to the rebels was initially seen as short sighted, but may have proven to be a useful tool to implement his reconciliation campaign. These improvements, however, leave major areas of uncertainty in dealing with terrorism, and with high unemployment and serious economic problems.

- **Morocco's difficult transition:** King Mohamed VI of Morocco has emerged as a more popular leader than his father and has eased much of the repression in the country. There are, however, so many demographic and economic problems in the country. His efforts at reform may, however, fail to bring stability. Many have emigrated and the lack of a popular base of support may hinder his push for further reforms. The Kingdom of Morocco also has major problems with terrorism, including the recent events in Casablanca and Rabat. It also continues to lack a middle class that is essential to its long-term economic, social, and political stability.

Given this list, there is no reason to assume that a broad ranging succession crisis will occur in the region, whether it is a matter of aging leaders or aging regimes. It would be equally unrealistic, however, to assume that all of these succession and leadership issues will be resolved peacefully and lead to stable new governments and will not affect the energy security of at least one Middle Eastern state between now and 2020.

At the same time, the MENA region is in a better shape politically than it was a decade ago. There is a trend toward openness and a willingness of the government to allow more open participations. This is reflected in a political dialogue that increasingly uses new terms like reform, transparency, and accountability. This is due to many factors, but the most obvious is the generational shift of the region. Many of the young leaders understand the social, economic, and political challenges and are more willing to increase the public participation in these matters.

The Next Class of Technocrats

Western risk analyses often focus on the "corruption" of MENA ruling elites, military and arms spending, and human rights issues. Practical governance and energy development, however, are heavily dependent on the quality of the region's technocrats. It is worth noting that the "technical" ministries, especially in the Gulf, such as oil, finance, economics, trade, etc., tend to be run by highly educated, and often Western educated, technocrats where the other important ministries tend to be held by members of the royal families. In fact, of the six Gulf countries, only Kuwait's oil minister is a member of its ruling family.

The region's technocrats have accomplished a great deal and have much to be proud of, but they have cumulatively caused as many problems as they have solved. Many MENA technocrats have not adapted rapidly enough to deal with the decline in state revenues and investment income. This may be least true in the energy sector, where the need for global competitiveness demands a high degree of competence. The problems still exist, however, and the problems in the energy sector cannot be separated from the overall management of the national budget.

Many have not learned to control costs, to properly plan and manage projects on the basis of realistic return of investment calculations, and often seek to create very large-scale projects as a basis for national and personal prestige. The end result has often been a vast drain on national budgets, a waste of resources on unprofitable ventures and investments, and a resistance to private investment and realistic national budgets and five-year plans.

A number of countries have also seen a drop in the Western education of their future technocrats without replacing it with high-quality domestic education. Local education overemphasizes rote learning, Islam, and politics in a way that means the pool of future talent with realistic public administration, economics, and business skills is too small to meet future needs.

Many countries are cutting their numbers of foreign managers and advisors or relegating them to less influential roles. The demands on the next generation of technocrats will also be much higher than in the past for one simple reason: there is far less money and surplus capital.

At the same time, the recent flood of wealth from energy exports provides the MENA region with an opportunity it missed following the first oil boom. The current generation and the upcoming generation of technocrats do seem to have learned many of the key lessons of the 1980s and 1990s, and they realize that they now have a second chance that many countries do not get. They have massive inflows of capital and a public that understands the real-world limitations of "oil wealth." The public is willing to see this new capital channeled into long-term strategic planning and into meaningful social, educational, and economic reforms. It is far from clear, however, how much this understanding will be translated into effective real-world actions.

THE ECONOMIC CHALLENGES

As mentioned earlier, for all the talk in the West about political reforms and the importance of "democratization," what the MENA countries need most is robust economic reform plans that are tailored to each country's needs, the strengthening of the rule of law, and the reforming of their educational systems to prepare the next generation for a globally completive economy especially given the youth explosion the MENA countries are facing.

Despite the inflow of oil wealth, the MENA region continues to lag behind most of the world's regions in economic growth, has high unemployment rates, suffers from large external debts, and has shown limited capacity to develop vibrant sectors. Some of these problems are regional; some are country specific and even sector specific.

During the last four years, there have been many attempts to report on the state of economic and human development in the region by the UN, the World Bank, the International Monetary Fund (IMF), and domestic agencies in each country. There has been a push toward major structural economic reforms, and many of the countries in the region have started on a campaign to reform their economies, but it remains uncertain how recent high oil prices and the newfound oil wealth may influence the pace of reforms, the governments' ability to deal with the unemployment problem, and the effort to diversify the economy.

Lagging Economic Growth

Economic development in the MENA region has been weak since the end of the oil boom in the late 1970s. MENA gross domestic product (GDP) growth rates have lagged behind other regions, and its growth patterns have been more volatile. In total GDP, the MENA region has the second lowest in the world, after the sub-Saharan African total. The MENA region's GDP rose from $199.79 billion in 1975 to $561.78 billion in 2004 (a 181-percent increase). This compares to East Asia's GDP growth from $256.17 billion in 1975 to $2,132 billion in 2004 (a 733-percent increase).[1]

If one observes the MENA region before the oil embargo of 1973, onto the boom years of the 1980s, the "oil crash" of the 1990s, and the rebound of 2004, the last three decades have been a volatile period for the MENA region, and these growth rates are reflective of the overall stagnation and stability in the region. Some experts have attempted to explain the lagging economic growth by comparing the MENA experience to other regions, others blamed it on corruption and the lack of "free market economies," while others blamed it on mismanagement of the economies. The obvious answer is that there are no obvious answers.

The recent influx of oil export income has helped, but has not reversed these trends. According to the World Bank, the real GDP of the MENA region grew at 7.31 percent in 1975, –0.19 percent in 1980, 4.07 percent in 1985, 6.71 percent in 1990, 2.99 percent in 1995, 3.57 percent in 2000, 5.23 percent in 2003, and 5.09 percent in 2004. While interregional comparison may be unfair, East Asia and the Pacific region outperformed the MENA region in growth rates: 13.16 percent in 1970, 6.74 percent in 1975, 7.14 percent in 1980, 7.48 percent in 1985, 5.57 percent in 1990, 9.59 percent in 1995, 7.29 percent in 2000, 8.18 percent in 2003, and 8.48 percent in 2004.[2]

Different countries in the MENA region have, however, had very different experiences. Some reports, such as the IMF and the World Bank, separate the countries in the MENA region depending on their dependence on natural resources or the abundance of cheap labor. Regardless of how they are categorized, the development of each country depends on a host of other factors that it is nearly impossible to understand the impact to oil or labor per se on the overall GDP growth. It is, however, worth noting that countries that have a high dependence on oil revenues tend to have higher volatility than those that do not.

The following figures show the quantitative trends in each of the MENA countries and compare them to the other regions of the world:

• Figure 3.1 shows the World Bank estimates of the MENA countries' real GDPs in constant 2000 dollars from 1970 to 2004. It covers the period of booms and crashes, and it highlights the general trend that the MENA countries, despite their oil wealth, have the

Figure 3.1   Real GDP of MENA Countries Compared to Other Regions: 1970–2004 (in 2000 Constant $Billions)

Country	1970	1975	1980	1985	1990	1995	2000	2003	2004
Algeria	19.22	25.46	34.36	43.47	45.15	45.73	53.45	60.98	64.15
Bahrain	...	...	3.98	3.71	4.65	6.45	7.97	9.37	...
Egypt	20.16	23.93	38.19	52.90	65.04	76.85	99.43	109.60	114.31
Iran	...	60.57	49.99	64.41	65.05	80.92	96.21	113.88	121.29
Israel	27.35	39.45	47.58	55.04	67.97	93.43	115.45	115.69	120.67
Jordan	...	2.03	4.22	5.44	5.14	7.25	8.45	9.66	10.38
Kuwait	30.77	27.40	28.99	22.90	...	35.67	37.02	40.11	...
Lebanon	...	...	...	...	8.44	15.00	16.59	18.67	19.85
Libya	12.37	9.36	14.71	9.59	...	...	34.50	40.63	42.46
Morocco	11.03	14.03	18.31	21.53	26.72	27.97	33.33	38.48	39.82
Oman	3.15	4.15	5.38	10.90	12.64	16.85	19.87	22.26	–
Saudi Arabia	44.07	110.02	153.72	121.76	144.13	166.00	188.44	204.25	214.94
Syria	3.42	6.37	8.80	10.16	10.93	16.02	18.04	19.73	20.44
Tunisia	4.24	6.35	8.63	10.60	12.26	14.83	19.47	21.91	23.17
UAE	–	23.27	48.47	42.29	47.53	52.57	70.25	...	...
Palestine	...	...	...	...	...	3.89	4.64	3.10	...
Yemen	...	...	...	...	5.51	7.22	9.44	10.58	10.86
MENA Total	–	199.79	239.64	300.53	325.61	390.73	470.77	534.56	561.78
East Asia & Pacific Total	189.30	256.17	358.28	511.63	736.17	1,189	1,599	1,965	2,132
Europe & Central Asia Total	...	...	...	...	1,048	828.68	952.68	1,076	1,153
L. America & Caribbean Total	734.44	980.84	1,271	1,304	1,426	1,693	1,983	2,010	2,130
South Asia Total	156.23	175.45	209.84	272.72	362.53	464.16	603.19	706.99	754.33
Sub-Saharan Africa Total	154.12	189.47	220.54	233.32	264.80	280.27	331.43	369.21	385.60

Source: World Bank, *World Bank Economic Indicators 2005,* available on-line.

lowest GDP as a region in the world. In 2004, the largest economy in the region was Saudi Arabia, followed by Israel, Iran, Egypt, and Algeria.

- Figure 3.2 shows the World Bank estimates of the MENA countries real GDP growth rates since 1970 through 2004. There are no consistent trends, which highlights the volatile nature of growth in the region. Kuwait is the most obvious case of volatility. It had negative growth rates throughout the 1970s and 1980s, but had high growth rates during the 1990s and into 2004.

- Figure 3.3 highlights the difference between highly oil-dependent economies and those that are not. It shows the IMF estimates of GDP growth rates of fuel exporters and those that are not. The quantitative trends are clear. The nonfuel exporter overperformed the fuel exporters and generally was less volatile. It is, however, worth noting here that even during 2005 when the price of oil was high, the overall growth, according to the IMF, is higher in the nonfuel exporters.

Figure 3.2 Real GDP Growth Rates of MENA Countries Compared to Other Regions: 1970–2004 (in % Growth Rates)

Country	1970	1975	1980	1985	1990	1995	2000	2003	2004
Algeria	8.86	5.05	0.79	3.70	0.80	3.80	2.40	6.80	5.20
Bahrain	...	...	...	–4.76	4.44	3.93	5.30	6.78	...
Egypt	5.60	8.94	10.01	6.60	5.70	4.67	5.11	3.20	4.30
Iran	...	5.45	–12.80	1.75	11.24	2.87	5.05	6.61	6.51
Iraq	3.16	10.49	–17.95	–14.33	–30.70	...	...	...	...
Israel	7.31	3.32	6.87	3.45	6.84	7.06	7.53	1.29	4.31
Jordan	...	...	19.01	3.46	0.97	6.19	4.10	3.98	7.46
Kuwait	3.18	–8.00	–20.62	–4.26	...	4.86	3.85	9.90	...
Lebanon	...	...	...	...	26.53	6.50	1.12	4.91	6.32
Libya	4.95	7.46	0.60	–8.84	...	...	1.15	9.14	4.50
Morocco	4.71	7.56	3.64	6.33	4.03	–6.58	0.96	5.24	3.49
Oman	13.90	24.43	6.04	14.01	7.51	3.20	5.40	2.47	...
Saudi Arabia	12.03	5.23	6.52	–4.32	8.33	0.20	4.86	7.66	5.23
Syria	–3.81	19.52	11.98	6.12	7.64	5.75	0.60	2.50	3.60
Tunisia	4.67	7.16	7.42	5.65	7.95	2.32	4.67	5.57	5.76
UAE	...	6.23	26.42	–6.57	17.53	6.12	12.30	...	...
Palestine	...	...	...	...	...	–2.42	–1.17	–1.67	...
Yemen	...	...	...	...	...	11.65	4.43	3.14	2.70
MENA Total	...	7.31	–0.19	4.07	6.71	2.99	3.57	5.23	5.09
East Asia & Pacific Total	13.16	6.74	7.14	7.48	5.57	9.59	7.29	8.18	8.48
Europe & Central Asia Total	...	...	...	...	–1.78	1.33	6.48	5.85	7.15
L. America & Caribbean Total	6.12	3.40	6.26	2.29	0.45	0.41	3.96	1.93	5.98
South Asia Total	5.81	7.03	6.42	5.67	5.63	6.96	4.21	7.80	6.70
Sub-Saharan Africa Total	7.67	1.13	4.23	1.34	1.07	3.82	3.42	4.27	4.44

Source: World Bank, *World Bank Economic Indicators 2005*, available on-line.

It is worth remembering that in spite of the differences shown in these figures, virtually all the major MENA oil producers were not prepared to deal with the oil crash of the 1990s. Most Middle Eastern states also suffer from economic mismanagement and excessive state control of their economies.

A few states like Kuwait, Qatar, and the United Arab Emirates have so much oil and gas wealth per capita that they can buy their way out of their mistakes in the short- to midterm, but even these three states need to reform their welfare and subsidy programs, reduce excessive state control, and reduce the "glitter factor" in their approach to their economic policies. Oil states like Algeria, Bahrain, Libya, Oman, Qatar, Saudi Arabia, and Yemen have very different levels of economic pressures, but all already need to move aggressively toward structural economic reform.

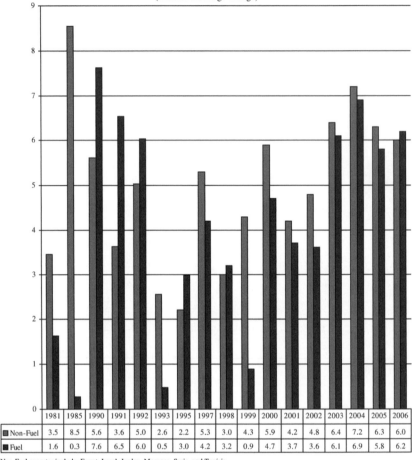

(Annual Percentage Change)

	1981	1985	1990	1991	1992	1993	1995	1997	1998	1999	2000	2001	2002	2003	2004	2005	2006
Non-Fuel	3.5	8.5	5.6	3.6	5.0	2.6	2.2	5.3	3.0	4.3	5.9	4.2	4.8	6.4	7.2	6.3	6.0
Fuel	1.6	0.3	7.6	6.5	6.0	0.5	3.0	4.2	3.2	0.9	4.7	3.7	3.6	6.1	6.9	5.8	6.2

Non-Fuel exporter include: Egypt, Israel, Jordan, Morocco, Syria, and Tunisia.
Fuel exporter include: Algeria, Bahrain, Iran, Kuwait, Oman, Qatar, Saudi Arabia, and the UAE.
Source: IMF, World Economic Outlook, various editions.

Figure 3.3   Real GDP Growth of Fuel Exporters vs. That of Nonfuel Exporters: 1981–2006

The other Middle Eastern states have uncertain near- to midterm economic prospects. The Palestinian economy was always weak and has been crippled by war. Egypt, Jordan, Lebanon, and Syria are all experiencing serious economic and demographic problems. The Iraqi economy is weak, suffers from major structural problems, and is plagued by uncertainty surrounding the insurgency. The Iranian economy is in near crisis, compounded by deep ideological conflicts over how to deal with the issue of reform.

The strategic, security, and political uncertainties affecting each country have also played a major role in determining the growth rates over the past 30 years. Much of the domestic capital in the MENA region left and went to investment in the region.

A lack of adequate local return on investment and the lessons from the fall of the Shah and the invasion of Kuwait helped drive this process. Large amounts of money left the MENA region, especially the Gulf countries, between the late 1970s and early 2000s. Estimates of this amount range from $400 billion to $1 trillion. This capital flight has slowed many necessary domestic investments into infrastructure and building small business enterprises and vibrant private sectors that can start to deal with the large unemployment problems.

In addition to a past lack of investment, many of the MENA economies have been driven by two main themes: government spending and private consumption. These are cyclical by nature for any economy, and they are more so in the MENA region because they are highly correlated with the price of oil. The 1990s were a downtime for many of the MENA countries because of the lack of government spending due to the high budget deficits, and the same was true for private consumption since the majority of the population was employed by the public sector that was highly dependent on the stability of the government and its oil revenues.

The present oil boom, and the fears of losing overseas investment growing out of 9/11, has helped reverse these trends. So has the structural economic reform that has begun in Algeria, Morocco, Tunisia, Egypt, Jordan, Saudi Arabia, Lebanon, Bahrain, and Oman. This reform, however, remains highly uncertain, and no country has yet carried out such reform to the point where it has a serious prospect of success. It remains equally uncertain how the high oil prices will influence the overall performance of the oil-dependent economies of the Gulf and North Africa.

The MENA region has also continued to suffer from a brain drain where many have immigrated to the West. In addition, the region's educational systems do not prepare students for a competitive world economy. These problems are compounded by high population growth rates, a sharply rising number of new entrants to the work force, and domestic unemployment problems.

Declining per Capita Income

The interregional differences between the MENA region and high economic growth regions are even more striking when one compares growth rates in GDP per capita. According to the World Bank, the MENA region's average GDP per capita was $1,406 in 1975 and $1,911 in 2004 (a 35-percent increase). In most of the Gulf countries, the GDP per capita actually decreased during the same period due to a rising population and declining oil prices. On the other hand, East Asia and the Pacific region's per capita income in 1975 was $204 and increased to $1,140 in 2004 (a 558-percent increase), and South Asia's per capita income increased from $219 in 1975 to $521 in 2004 (a 137-percent increase).[3]

Looking at year-by-year GDP per capita growth rates, the MENA region's GDP per capita grew by 4.41 percent in 1975, −3.15 percent in 1980, 0.87 percent in 1985, 3.17 percent in 1990, 0.90 percent in 1995, 1.69 percent in 2000, 3.36 percent in 2003, and 3.34 percent in 2004. The major MENA countries also saw high rates of fluctuations in growth during the last three decades. For example, Saudi

Arabia's GDP per capita, in constant 2000 dollars, increased from $7,672 in 1970, to $16,402 in 1980, but started to decline to $9,120 in 1990 and $9,093 in 2000.

It is important to note that estimates of GDP per capita differ by source because they depend on a number of factors that are hard to measure. For example, they depend on knowing the economically active population and the actual GDP. In many MENA countries, even these "basic" statistics are hard to measure.

Even so, the World Bank provides a good benchmark for all the regions by using the same method of estimation for all countries. Figures 3.4 and 3.5 compare the MENA countries' GDP per capita levels to those of other regions.

Figure 3.4 shows the trends of the GDP per capita and an overall decline in MENA country income since the oil booms of the 1970s. It shows that even some

Figure 3.4   GDP per Capita by MENA Country Compared to Other Regions: 1970–2004 (in 2000 Constant $)

Country	1970	1975	1980	1985	1990	1995	2000	2003	2004
Algeria	1,398	1,589	1,841	1,987	1,804	1,630	1,759	1,916	1,981
Bahrain	. . .	. . .	11,926	8,727	9,238	11,183	11,897	13,168	. . .
Egypt	610	659	934	1,137	1,240	1,321	1,554	1,622	1,663
Iran	. . .	1,824	1,278	1,367	1,196	1,373	1,511	1,715	1,812
Israel	9,198	11,418	12,270	13,002	14,585	16,850	18,358	17,298	17,752
Jordan	. . .	1,122	1,937	2,059	1,622	1,728	1,729	1,819	1,908
Kuwait	41,357	27,206	21,082	13,374	. . .	19,796	16,906	16,738	. . .
Lebanon	. . .	. . .	. . .	. . .	2,322	3,746	3,834	4,151	4,358
Libya	6,228	3,826	4,836	2,533	. . .	. . .	6,587	7,308	7,483
Morocco	720	811	945	994	1,111	1,060	1,161	1,278	1,302
Oman	4,351	4,896	4,886	7,806	7,771	7,892	8,244	8,565	. . .
Saudi Arabia	7,672	15,173	16,402	9,836	9,120	9,118	9,093	9,066	9,259
Syria	546	856	1,011	978	902	1,127	1,115	1,135	1,150
Tunisia	827	1,132	1,353	1,460	1,503	1,655	2,036	2,214	2,315
UAE	. . .	46,079	46,473	30,667	26,809	21,805	21,636	. . .	. . .
Palestine	. . .	. . .	. . .	. . .	. . .	1,625	1,563	920	. . .
Yemen	. . .	. . .	. . .	. . .	464	475	539	552	550
MENA Total	. . .	1,406	1,458	1,568	1,471	1,570	1,720	1,849	1,911
East Asia & Pacific Total	169	204	264	348	461	696	886	1,060	1,140
Europe & Central Asia Total	. . .	. . .	. . .	. . .	2,249	1,754	2,010	2,283	2,443
L. America & Caribbean Total	2,615	3,088	3,565	3,293	3,281	3,571	3,880	3,765	3,935
South Asia Total	219	219	233	271	324	376	445	496	521
Sub-Saharan Africa Total	533	572	575	527	519	483	504	524	536

Source: World Bank, *World Bank Economic Indicators 2005,* available on-line.

Figure 3.5   GDP per Capita Growth Rates by MENA Country Compared to Other Regions: 1970–2004 (Annual Percentage Change)

Country	1970	1975	1980	1985	1990	1995	2000	2003	2004
Algeria	5.65	1.87	–2.26	0.50	–1.75	1.71	0.93	5.08	3.45
Bahrain	...	...	...	–8.72	1.53	0.33	3.10	4.71	...
Egypt	3.41	6.75	7.27	3.89	3.31	2.66	3.13	1.39	2.51
Iran	...	2.18	–15.77	–2.18	8.84	1.23	3.51	5.24	5.66
Iraq	–0.12	6.90	–20.61	–17.13	–32.9	...	...	...	...
Israel	3.81	0.99	4.34	1.64	3.58	4.24	4.73	–0.56	2.62
Jordan	...	...	14.55	–0.17	–2.66	2.79	0.97	1.30	4.85
Kuwait	–4.27	–13.50	–25.18	–7.42	...	...	–0.08	7.08	...
Lebanon	...	...	...	...	24.07	4.50	–0.20	3.60	5.01
Libya	0.74	2.98	–3.88	–12.41	...	...	–0.79	6.96	2.39
Morocco	1.96	5.06	1.35	4.03	1.93	–8.21	–0.69	3.59	1.89
Oman	10.82	19.39	0.49	9.79	4.16	0.35	2.69	0.07	...
Palestine	...	...	...	...	...	–6.18	–5.31	–5.63	...
Saudi Arabia	7.58	0.20	0.81	–9.29	3.61	–2.39	2.21	4.59	2.12
Syria	–6.99	15.65	8.31	2.40	4.11	2.50	–1.87	0.15	1.28
Tunisia	2.71	4.92	4.59	2.48	5.36	0.70	3.49	4.35	4.53
UAE	...	–11.15	14.58	–14.09	11.17	–2.42	4.90	...	...
Yemen	...	...	...	...	...	8.56	1.50	0.07	–0.37
MENA Total	–	4.41	–3.15	0.87	3.15	0.90	1.69	3.36	3.34
East Asia & Pacific Total	10.15	4.70	5.55	5.84	3.88	8.24	6.30	7.27	7.59
Europe & Central Asia Total	...	...	...	...	–2.39	0.44	6.53	5.77	7.01
L. America & Caribbean Total	3.48	0.94	3.94	0.26	–1.32	–1.24	2.49	0.50	4.51
South Asia Total	3.32	4.52	3.99	3.47	3.44	4.93	2.36	6.04	5.01
Sub-Saharan Africa Total	4.88	–1.63	1.10	–1.54	–1.78	1.01	1.09	1.98	2.35

Source: World Bank, *World Bank Economic Indicators 2005,* available on-line.

oil-producing countries experienced a decline in their overall per capita income in constant U.S. dollars. On the other hand, the non-oil-producing nations such as Bahrain, Egypt, Israel, Jordan, Morocco, Tunisia, and even Oman have seen some improvement in the level of per capita incomes.

Figure 3.5, on the other hand, compares the growth rates for each of the countries and compares them to other regions of the world. It shows the high level of volatility and the uncertainty surrounding economic stability in the region.

High oil revenues should improve this situation, at least in the short-term. MENA countries are projected to grow their per capita income by 5.2 percent between 2000 and 2006, and by 2.4 percent between 2006 and 2015, as shown in Figure 3.6. These numbers are the World Bank's 2005 estimates and seem to be based on the

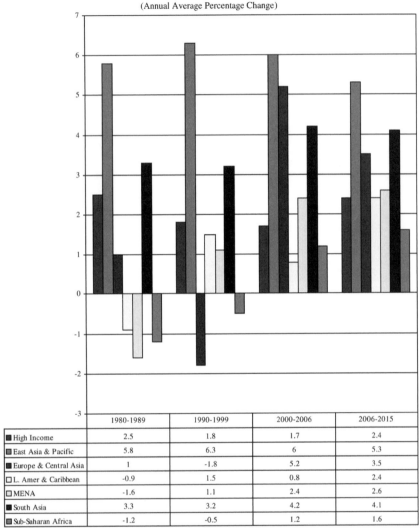

(Annual Average Percentage Change)

	1980-1989	1990-1999	2000-2006	2006-2015
■ High Income	2.5	1.8	1.7	2.4
▨ East Asia & Pacific	5.8	6.3	6	5.3
■ Europe & Central Asia	1	-1.8	5.2	3.5
☐ L. Amer & Caribbean	-0.9	1.5	0.8	2.4
☐ MENA	-1.6	1.1	2.4	2.6
■ South Asia	3.3	3.2	4.2	4.1
▨ Sub-Saharan Africa	-1.2	-0.5	1.2	1.6

Source: World Bank, Global Economic Prospects 2005.

Figure 3.6   World Bank Regional Average Estimates for GDP per Capita Growth: 1980–2015

assumption that oil prices will decrease in the near future—which may or may not hold true. However, other developing regions are still projected to outperform the MENA region. For example, East Asian per capita GDP is estimated to increase by 6.0 percent in 2000–2006 and by 5.3 percent between 2006 and 2015. South Asia is estimated to have similar growth rates during the same period: 4.2 percent in 2000–2006 and 4.1 percent in 2006–2015.

Lack of Inflationary Tendencies (Paradox)

This mixed picture of GDP and per capita income growth is compounded by a phenomenon common to most MENA economies. They all have surprisingly low inflation rates. There are obviously exceptions, such as Lebanon and Israel's hyperinflations in the 1980s, Egypt's relatively high inflation rates, and Libya's 1970–1995 high inflation rates. Across the Gulf and North Africa, the MENA region has enjoyed no inflationary tendencies.

Figure 3.7 shows the World Bank estimate of the MENA country inflation rates between 1980 and 2005. There are two major trends: First, the rates are generally low and sometimes negative, especially during "the oil crash" of the 1990s. Second, the rates are volatile, despite their low level, and especially for the oil-producing states.

In the countries with pegged exchange rates and high dependence on oil exports [the Gulf Cooperation Council (GCC) states], some experts argue that more of the growth is due to growth in export prices, which do not have much direct influence on the local consumer prices. That is true, but it is equally true to note that there is no meaningful "real economic growth," and that generally translates into low domestic prices.

This paradox has become more striking as oil revenues have risen sharply. The MENA region, especially the Gulf, is experiencing a tremendous economic boom. Some believe that this boom has all the signs of a bubble due to the high oil prices and speculation. Others believe that the "petrodollars" are yet to come into the economy, and most of this growth is due to real market forces based on sound economic fundamentals.

Figure 3.7   MENA Inflation Rates by Country: 1980–2005 (Annual Percentage Change in Consumer Prices)

Country	1980	1985	1990	1995	2000	2001	2002	2003	2004	2005
Algeria	9.7	10.4	9.3	29.8	0.3	4.2	1.4	2.6	5.4	4.5
Bahrain	3.8	−2.4	−0.3	3.1	−3.6	−1.2	−0.5	0.6	1	1.2
Egypt	20.5	12.1	22.2	9.4	2.8	2.4	2.4	3.2	5.2	5.7
Iran	20.6	4.4	9	49.4	12.6	11.4	15.8	15.6	15.6	15
Israel	130.9	308.8	17.1	10	1.1	1.1	5.7	0.7	−0.3	1.4
Jordan	11	2.9	−3.3	2.3	0.7	1.8	1.8	2.3	3.5	1.8
Kuwait	6.9	1.4	75.2	2.7	1.8	1.7	1.4	1.2	1.7	1.6
Lebanon	23.9	69.4	88.9	10.3	−0.4	−0.4	1.8	1.3	3	2
Libya	14.3	9.1	8.5	8.3	−2.9	−8.8	−9.9	−2.1	2.1	3
Morocco	9.4	7.7	6	6.1	1.9	0.6	2.8	1.2	2	2
Oman	10	−4	10	−1.1	−1.2	−1.1	−0.6	−0.4	1	0.7
Qatar	6.8	1.1	4.5	3	1.7	1.4	1	2.3	3.5	3
Saudi Arabia	4.4	−3.1	−1	5	−0.6	−0.8	−0.6	0.5	2.5	0.8
Tunisia	10	7.6	6.5	6.2	3	1.9	2.8	2.8	3.4	2.7
UAE	10.1	3.5	0.6	4.4	1.4	2.8	3.1	2.8	3.4	2.1
Yemen	N/A	N/A	33.5	62.5	10.9	11.9	12.2	10.8	15.3	...

Source: IMF, World Economic Outlook Database, September 2004. N/A = not available.

Regardless of the prognosis, the difference between nominal and real GDP growth tends to be in the double digits, while inflation rates sometimes are less than 1 percent. In 2005, the MENA inflation rates ranged from 0.7 to 5.7 percent. Given the projected growth rates of at least 5 percent for most of the oil-producing nations, the inflation rates remain low. The impact of this phenomenon on the fiscal balance, the overall health of the capital markets, and consumer confidence remain unclear. The reporting on the current "boom" is just starting.

Fiscal Balance and the "Overspending"

Budget deficits have also been a major problem, at least in the past. The World Bank estimates that the MENA states ran a budget deficit that was approximately 4.8 percent of their GDP between 1990 and 2000. The largest was in Kuwait which had a 35-percent budget deficit/GDP ratio. This is, however, due to the large spending following the 1990 invasion by Iraq and the subsequent rebuilding. Iran and Oman also ran large budget deficits compared to their GDP during the same period, −6.1 percent and −8.0 percent, respectively. Saudi Arabia ran a −2.4 percent of GDP budget deficit. Algeria, the United Arab Emirates, and Morocco were the only states in the MENA region that ran budget surpluses.[4]

Most of the larger MENA countries have been running such deficits since the 1980s. Much of this spending went into "welfare" programs and defense spending. There has also been mismanagement of national budgets. According to the U.S. State Department, the Middle Eastern countries' military spending represented 32.4 percent of government expenditures in 1985, 47.1 percent in 1990, 22.3 percent in 1995, and 21.4 percent in 1999. The North African countries spent less: 19.5 percent in 1985, 11.9 percent in 1995, and 13.1 percent in 1999. These numbers are much larger than those for any other region of the world. NATO, for example, spent 14.4 percent in 1985, 9.7 percent in 1995, and 9.1 percent in 1999 of total military expenditures.[5]

Defense spending, however, is not the only strain on the budget. The overall civil spending is also high. Many of the MENA countries embarked on large welfare and entitlement programs in the 1970s and the 1980s. The oil-producing states in the MENA region built their economies based on free health care, education, and no taxes. Most of these economies cannot support themselves without the oil export revenues, and when the oil prices are low the overall budget balance suffers.

Once again, high oil revenues have eased the situation. Figure 3.8 shows the World Bank estimates between 1990 and 2006. It shows that, on average, the MENA states started running budget balances between 2002 and 2005. The World Bank estimates that the MENA region had a budget balance-to-GPD ratio of 0.30 percent in 2002, 2.2 percent in 2003, 3.2 percent in 2004, and 1.6 percent in 2005. Kuwait again leads the way as having the highest ratio of 21.2 percent in 2002, 24.2 percent in 2003, 23.7 percent in 2004, and 17.5 percent in 2005. The other Gulf States, such as the United Arab Emirates, continue to have close to a balanced budget with a small surplus in the last four years. Saudi Arabia is also experiencing large budget

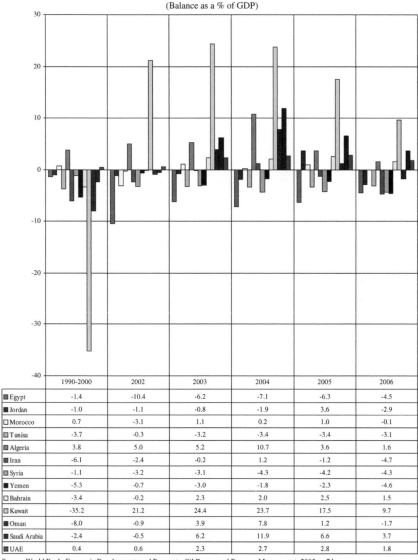

(Balance as a % of GDP)

	1990-2000	2002	2003	2004	2005	2006
▣ Egypt	-1.4	-10.4	-6.2	-7.1	-6.3	-4.5
■ Jordan	-1.0	-1.1	-0.8	-1.9	3.6	-2.9
▢ Morocco	0.7	-3.1	1.1	0.2	1.0	-0.1
▨ Tunisa	-3.7	-0.3	-3.2	-3.4	-3.4	-3.1
▤ Algeria	3.8	5.0	5.2	10.7	3.6	1.6
■ Iran	-6.1	-2.4	-0.2	1.2	-1.2	-4.7
▣ Syria	-1.1	-3.2	-3.1	-4.3	-4.2	-4.3
■ Yemen	-5.3	-0.7	-3.0	-1.8	-2.3	-4.6
▢ Bahrain	-3.4	-0.2	2.3	2.0	2.5	1.5
▢ Kuwait	-35.2	21.2	24.4	23.7	17.5	9.7
■ Oman	-8.0	-0.9	3.9	7.8	1.2	-1.7
■ Saudi Arabia	-2.4	-0.5	6.2	11.9	6.6	3.7
▣ UAE	0.4	0.6	2.3	2.7	2.8	1.8

Source: World Bank, Economic Developments and Prospects: Oil Booms and Revenue Management, 2005, p. 74.

Figure 3.8   Estimates of MENA Fiscal Balance Relative to GDP by Country: 1990–2006

surpluses, but Iran is not experiencing the same level of surpluses like the other Gulf countries. The nonoil producers such as Syria and Yemen, however, have continued to suffer from large budget deficits.

Surges in oil revenues, however, are scarcely a substitute for effective fiscal management and economic planning. Figure 3.9 shows the IMF estimates of the impact of

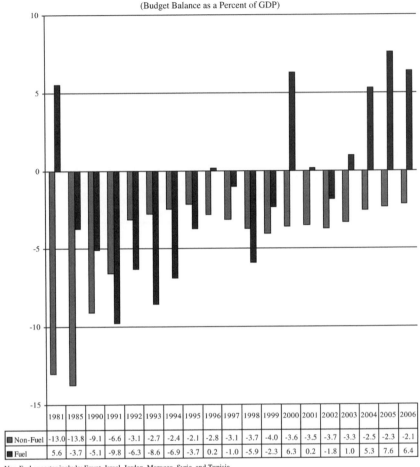

(Budget Balance as a Percent of GDP)

	1981	1985	1990	1991	1992	1993	1994	1995	1996	1997	1998	1999	2000	2001	2002	2003	2004	2005	2006
■ Non-Fuel	-13.0	-13.8	-9.1	-6.6	-3.1	-2.7	-2.4	-2.1	-2.8	-3.1	-3.7	-4.0	-3.6	-3.5	-3.7	-3.3	-2.5	-2.3	-2.1
■ Fuel	5.6	-3.7	-5.1	-9.8	-6.3	-8.6	-6.9	-3.7	0.2	-1.0	-5.9	-2.3	6.3	0.2	-1.8	1.0	5.3	7.6	6.4

Non-Fuel exporter include: Egypt, Israel, Jordan, Morocco, Syria, and Tunisia.
Fuel exporter include: Algeria, Bahrain, Iran, Kuwait, Oman, Qatar, Saudi Arabia, and the UAE.
Source: IMF, World Economic Outlook, various editions.

Figure 3.9   Fiscal Balance of Fuel Exporters vs. That of Nonfuel Exporters: 1981–2006

oil dependence on the overall fiscal balance. It is all too clear that most of the MENA region countries had deficits since the 1980s, and they are just starting to achieve surpluses following the recent hike in oil prices.

It is equally clear that the fiscal balance of the MENA oil-producing states is more volatile than that of the states that are not fuel exporters. According to the IMF, the nonfuel exporters' budget balance was –13.0 percent (deficit) of GDP in 1981, –6.6 percent in 1991, –3.5 percent in 2001, and it projects –2.1 percent in 2006. Compared to the fuel producers, in 1981 their budget balance was 5.6 percent (surplus) of GDP, –9.8 percent in 1991, 0.2 percent in 2001, and 7.6 percent in 2005.

In short, much of the future economic stability of MENA oil producers will depend on high oil revenues and whether they use their second chance wisely. Social and internal security pressures may force them to overspend in the wrong areas when they need to spend on real growth, economic diversity, and job creation. They also need to spend to revitalize many important infrastructures that have aged since they were built in the 1970s and 1980s.

It is too early to call whether these recent surpluses will be mismanaged. Recent indications show that they may be repeating their glitter factor spending on "dream" projects such as man-made islands and theme parks in the middle of the desert, especially in the smaller Gulf States. There are far more important uses of state revenues.

Trade, Oil Wealth, and Oil Nonwealth

The good news is that many MENA oil states are starting to liberalize their economies; the bad news is that they are doing so at rates that may not be sustainable. Many in the Gulf are signing trade deals with the United States and other major trading partners. This has required many of the countries in the region to change their laws and regulations to meet the requirements of these deals. The overall impact of these deals may not be seen for many years. It remains equally uncertain what the impact of these liberalization policies will be on the overall economic diversification campaigns.

Figure 3.10 shows the CIA *World Fact Book 2005* estimates of MENA country external debt. Iraq has the largest debt of $125 billion, followed by Israel with $74.46 billion, Saudi Arabia with $34.35 billion, Egypt with $33.75 billion, Algeria with $21.90 billion, Qatar with $18.62 billion, Morocco with $17.07 billion, Lebanon with $15.84 billion, Kuwait with $15.02 billion, Iran with $13.40 billion, Jordan with $7.32 billion, Bahrain with $6.22 billion, United Arab Emirates with $5.90 billion, Yemen with $4.81 billion, Oman with $4.81 billion, Libya with $4.07 billion, Syria with $4.00 billion, and West Bank and Gaza with $0.11 billion each.

The world's dependence on MENA energy exports often disguises the fact that the overall role of energy exports has shrunk just as steadily as a percentage of world trade as its share of the world's GDP. In broad terms, MENA trade has shrunk as a share of world exports for nearly half a century. The only exceptions are a few years like 1974, 1980, and 2005—when sudden massive peaks occurred in the value of oil exports.

This drop in the global importance of MENA trade reflects a drop in overall regional competitiveness relative to the industrialized world, Asia, and Latin America. It also reflects the inefficiency of state industries and MENA financial systems, a comparative inability to attract foreign investment, and a growing dependence on imports from other regions. MENA nations produce less and less goods and services that are competitive within each country or on an intraregional basis. All MENA nations now trade primarily with trading partners outside the region, and efforts to break down regional trade barriers can accomplish little at best.

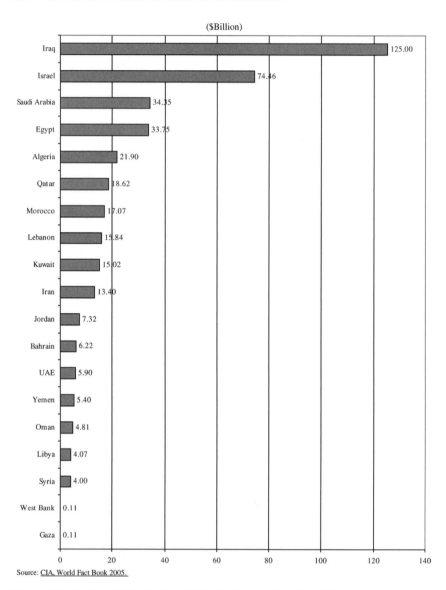

Figure 3.10   External Debt for the MENA Region by Country: 2005

MENA countries are especially dependent on trade for two main reasons: First, they are highly dependent on oil revenues exports, and oil represents at least a third of their GDP. Second, the MENA countries are food importers, and most of their basic consumption comes from the outside world. Many of these countries have their largest trading partners in the United States, Japan, and the European Union.

Figure 3.11 shows the World Bank estimates of the share of trade in GDP. It shows the current account balance as a percentage of GDP in 2004: The MENA region's current account was 14.40 percent, East Asia and the Pacific region was 1.5 percent, sub-Saharan Africa was 1.30 percent, Latin America and the Caribbean was 0.70 percent, Europe and Central Asia was 0.10 percent, South Asia was –0.50 percent, and the high income was –0.80 percent.

This high dependence on trade highlights the importance of imports and oil export revenues to the economies of the region, and the overall impact of current account balance on the growth of the MENA region. Figure 3.12 shows the World Bank estimates of the current account balance in current billions of U.S. dollars between 1990 and 2006. Saudi Arabia, Iran, Kuwait, and United Arab Emirates have the higher surpluses. This is also a reflection of their importance as energy exporters. The low figures are for negligible exporters like Yemen, Jordan, and Syria.

These trends are likely to remain for the foreseeable future. Figure 3.13 shows the volume of exports and imports for the MENA region in 2005. These volumes were highly influenced by the hikes in oil prices that happened in late 2005. The figure

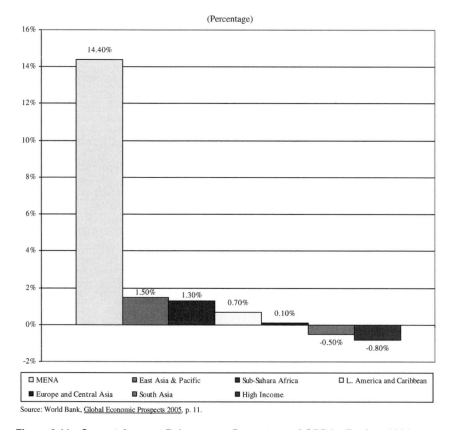

Source: World Bank, Global Economic Prospects 2005, p. 11.

Figure 3.11   Current Account Balance as a Percentage of GDP by Region: 2004

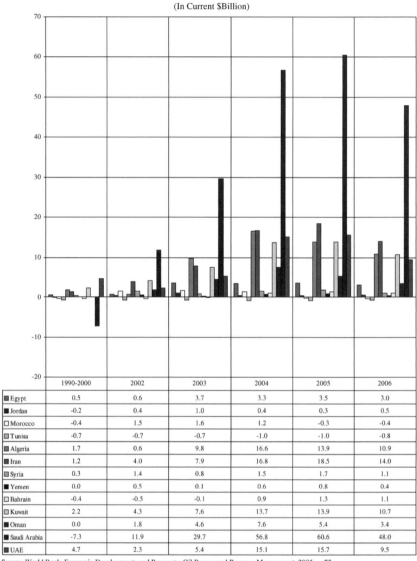

(In Current $Billion)

	1990-2000	2002	2003	2004	2005	2006
◼ Egypt	0.5	0.6	3.7	3.3	3.5	3.0
◼ Jordan	-0.2	0.4	1.0	0.4	0.3	0.5
☐ Morocco	-0.4	1.5	1.6	1.2	-0.3	-0.4
▨ Tunisa	-0.7	-0.7	-0.7	-1.0	-1.0	-0.8
◼ Algeria	1.7	0.6	9.8	16.6	13.9	10.9
◼ Iran	1.2	4.0	7.9	16.8	18.5	14.0
▨ Syria	0.3	1.4	0.8	1.5	1.7	1.1
◼ Yemen	0.0	0.5	0.1	0.6	0.8	0.4
☐ Bahrain	-0.4	-0.5	-0.1	0.9	1.3	1.1
☐ Kuwait	2.2	4.3	7.6	13.7	13.9	10.7
◼ Oman	0.0	1.8	4.6	7.6	5.4	3.4
◼ Saudi Arabia	-7.3	11.9	29.7	56.8	60.6	48.0
◼ UAE	4.7	2.3	5.4	15.1	15.7	9.5

Source: World Bank, Economic Developments and Prospects: Oil Booms and Revenue Management, 2005, p. 73.

Figure 3.12   Estimates of MENA Current Balance by Country: 1990–2006

also highlights the difference between oil producers and nonoil producers in the region.

For example, Saudi Arabia exported $113 billion and imported $36.21 billion compared to Egypt, which exported $11.0 billion and imported $19.21 billion, and Israel which exported $34.341 billion and imported $36.84 billion during the same year. They are based on the CIA *World Fact Book 2005,* and they were estimated

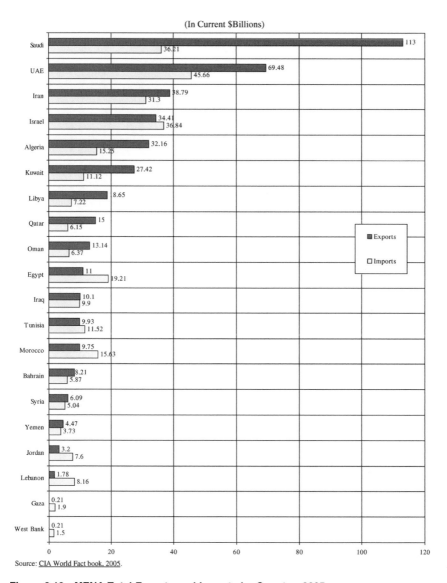

(In Current $Billions)

Figure 3.13   MENA Total Exports and Imports by Country: 2005

in early July 2005—which does not take into account the high oil prices following Hurricanes Katrina and Rita and the hikes in oil production by Saudi Arabia.

The Stock Market Bubbles: Signs of Overvaluation

The growth in oil export revenues has also led to a situation where the Gulf States and many other states in the MENA region are experiencing a boom in their stock markets and domestic investment. The Saudi stock market *Tadawul* index increased by roughly 540 percent during the last five years. From June 2001 to October 2005, the Kuwaiti stock exchange rose by roughly 560 percent, and from October 2002 to October 2005, the Dubai stock exchange increased by approximately 1,024 percent.[6]

It is important to note that the market cap of each stock market represents a large percentage of GDP. They range from as low as 18 percent in Oman to as high as 143 percent in Kuwait. These numbers show how important the stock markets in the Gulf States have been to the recent high growth rates in their respective economies. It is equally important to note that the GCC stock markets outperformed their 2003–2004 growth rates in 2004–2005.

In addition, according to Al-Rajhi Bank, market capitalization of the GCC stock markets increased by 92 percent between September and January 2005. It reached $1.042 trillion on September 29, 2005, compared to $543 billion on December 31, 2004, and $119 billion in December 2000. Saudi Arabia contributed to 55.8 percent of the total increase, with the remainder accounted for by Dubai (21.1 percent), Kuwait (10.8 percent), Qatar (9.5 percent), Bahrain (1.6 percent), and Oman (1.2 percent).[7]

The Gulf stock markets have seen major initial public offerings (IPOs) such as Saudi Telecom Company (STC) and Dana Gas of the United Arab Emirates. In each case, investors oversubscribed the shares of these companies. For example, Dana Gas of the United Arab Emirates was oversubscribed by 140 percent on its initial public offering in early October 2005. According to the Saudi American Bank (SAMBA), in recent Saudi IPOs, STC was 2.4 times oversubscribed, Saudi Sahara petrochemicals was 124 times oversubscribed, and Ettihad Ettisalat was 50 times oversubscribed.[8]

In addition, the GCC stock markets have recently had much higher than normal price-earning (PE) ratios. For the first quarter of 2005, the Saudi stock market had a PE ratio of roughly 39. In March 2005, the PE of the UAE stock market reached as high as 47. It is significant to note that "sectoral" PE ratios are even higher. According to SAMBA, the PE ratio for the first quarter of 2005 in the agricultural sector was 88.6, electricity was 71.3, service was 54.2, and insurance was 34.2. It is also equally important to compare these numbers to roughly a PE ratio of 20 for the Dow Jones.[9]

Many experts question whether this growth is due to real market forces or the sign of "irrational exuberance" on the part of Gulf and international investors. As is the case with many intricate economic systems, the growth in the GCC stock markets is due to an array of factors.

There are no measures or warnings of economic or stock market bubbles. Few people—if any—predicted the U.S. stock market bubble in the late 1990s or the Asian economic meltdown of 1997. Bubbles, however, do eventually burst. The timing, the amount of loss, and the shape of recovery, however, are often uncertain.

The apparent overvaluation of the GCC stock markets—represented by high PE ratios and unprecedented growth in market capitalization—present a worrisome trend. An active monetary policy to tighten the money supply and cool off the economies in the Gulf may be necessary in the short-term. If this bubble is left to burst on its own, it may have long-term economic and strategic implications.

The consequences of such a market correction are hard to predict, due to the lack of historical data for the Gulf stock markets. Capital market laws and regulations are just beginning, economies in the Gulf are just starting to open up, and the health of the Gulf economies depend largely on the volatility of oil prices.

Understanding the underlying forces at play is of enormous importance in crafting a policy to address these trends. It is equally important that this be done before beginning any efforts to launch a common currency in the GCC, which is currently scheduled for 2010, and before any further efforts to open up the Gulf economies through trade deals.

Reasons for the Exuberance

The Gulf economies may be different than others that have experienced economic and stock market bubbles, but this seems unlikely. With sustained high oil prices predicted for the foreseeable future compounded by the high dependence of the Gulf economies on oil, the GCC states will likely sustain moderate to high growth rates in the near future. The problem is that current growth in the GCC markets is nominal, and there is little evidence that this growth is based on real structural economic changes. Several things happened during the last five years that made this growth possible.

- First, the Gulf economies, despite efforts to diversify, are still highly dependent on the oil and petrochemical sectors. For most of the GCC economies, oil revenues account for roughly a third of the GDP, as much as 75 percent of the budget, and approximately 90 percent of export revenues. During the last five years, the price per barrel of oil increased by roughly 108 percent. In addition to high oil prices, global demand for oil increased from 78.0 million barrels per day (MMBD) in 2001 to 82.4 MMBD in 2004. According to the Energy Information Administration (EIA), during the same period, the Gulf production capacity is estimated to have increased from 14.1 MMBD in 2001 to 17.26 MMBD, matching nearly three-quarters of the total world demand increase.

  The EIA estimates that Saudi oil export revenues, in constant 2000 dollars, were $59.64 billion in 2001 compared to $108.03 billion in 2004, representing roughly an 81-percent increase. The United Arab Emirate's oil export revenues in 2001 were $18.03 billion compared to $28.27 billion in 2004 (a 57-percent increase). Kuwait earned $18.63 billion in 2001 and $25.62 billion in 2004 (a 37-percent increase). Qatar's oil revenues rose from $7.03 billion in 2001 to $12.64 billion in 2004 (an 80-percent increase). Bahrain is not a major oil producer, and it is worth noting that Bahrain's stock market performed the worst out of the seven GCC markets.[10]

- Second, due to high budget surpluses, government spending in the Gulf reached all time highs. Saudi Arabia declared its intention of spending its budget surplus on rebuilding infrastructure, repaying its public debt, and modernizing its educational systems. For example, in 2004, the government announced that the $26.1-billion surplus would be spent on two broad areas: $15.2 billion would be used to pay down the Kingdom's public debt and the rest would go toward modernizing infrastructure. In addition, the Kingdom has announced a series of major energy projects to increase its production capacity to 12.5 MMBD by 2009, which will cost the government an estimated $16.5 billion. Other Gulf States have also embarked on similar spending patterns.

- Third, there has been repatriation of Gulf capital from the West. Large amounts of Arab investments were moved from the United States and Europe to the region out of fear of it being frozen after new regulations following the attacks of 9/11, as a backlash against the U.S. invasion of Iraq, and due to general Arab anger at the U.S. position on the Palestinian *Intifadhah*. The data on the total amount of Gulf capital in the West remains uncertain—estimates range from $400 billion to $800 billion. By all accounts a large portion of this capital has been repatriated and invested domestically in the GCC.

- Fourth, the Gulf has seen tremendous growth in its access to information during the last five years. Investors have the ability to do research on individual companies, as well as the global economy, and the ability to trade online at cheaper prices. The access to information technology (IT), however, has also complicated regulatory agencies' ability to monitor the capital markets. Rumors and "hyping" of stocks on the Internet are commonplace, and it is a near impossibility for the young regulatory agencies to control it.

- Fifth, the GCC countries are "liberalizing" their economies. During the last year, Bahrain signed a Free Trade Agreement (FTA) with the United States, and Saudi Arabia has signed a deal with the United States, which paved the way for its accession to the World Trade Organization in December 2005. In addition, Oman and the United Arab Emirates are negotiating FTAs with the United States, and Qatar and Kuwait are likely to follow suit. In order to qualify for trade deals, the GCC countries have opened up their capital markets to outside investors, introduced foreign capital laws, and streamlined investment inflow. It is, however, important to note that many of these reforms are just starting to take effect and their long-term effect is yet to be known, but they have at least given the perception of diversification.

- Sixth, there have been signs of wealth distribution that did not exist in the Gulf States before. For example, according to SAMBA, more than half of the Saudi population subscribed in the Al-Bilad Bank's IPO in early 2005. This can be attributed to the emergence of a new generation of Gulf investors, who tend to be younger, come from a diverse educational and economic background, and are more willing to take risks. That is not to say that the entire body of investors is young, but this generational shift is important and, given the dynamics of the demographics, this will continue for decades to come.

- Seventh, the growths in the GCC stock indices have been driven by giant companies such as STC, Dana Gas, and Al-Bilad Bank. According to a recent SAMBA report, 45 percent of the Saudi stock market capitalization comes from three companies: Saudi Arabian Basic Industries Co., Saudi Electricity Co., and Saudi Telecom Co. (STC). This is an indication of Saudi and GCC efforts to privatize major areas of their economies such as communication, electricity, and transportation.[11]

These are important factors that provide a broad picture of what took place in the GCC economies during the last decade. They are factors that can also have effects that extend beyond the economies of the Gulf, and the uncertainty surrounding the GCC stock markets can add to an "instability premium" in the oil market, international security, and the global economy.

Short- and Medium-Term Capital Market Risks

The key questions that these issues raise for Gulf stability are, do these issues call for state intervention or should it be left to market forces? And, what are the risks of inaction? There are no simple answers to these important questions. It is all too easy to recommend active policy to burst the bubble before it becomes unmanageable. It is much more difficult, however, to craft sensible economic policies to cool the economy off without impeding real economic progress.

An economic meltdown in the Gulf can have dire consequences for the global energy market and stability of some of these countries. The following are key areas of uncertainty in the GCC capital markets, which also apply to the state of capital markets in the other MENA states:

- Transparency in the banking systems in the Gulf States have improved over the last few years, and that has added to confidence in the capital markets. Many analysts, however, believe that profits announced by some banks in the GCC, particularly in the United Arab Emirates, may not be sustainable in the long-run, especially given the fact that banks represent at least half of the listed stocks on the Gulf exchanges.

- Currencies in the Gulf States are pegged to the U.S. dollar. No one fully knows their "real" value in the long-term or their level of volatility. They have always been pegged, and it is important that countries in the GCC do not float them without careful study of their foreign currency reserves, and their ability to support a reasonable value in the short- to medium-terms.

- While there have been some efforts by the Gulf States to provide a level of transparency on the flow of capital, there is limited data to suggest that countries in the region have developed the necessary mechanisms to deal with excess capital from the oil boom and the repatriation of capital from the West.

- The Gulf States are trying to build an atmosphere that encourages entrepreneurship and builds vibrant private sectors. In light of the high capital inflows, the GCC countries should create venture capital funds to channel some of this excess capital into meaningful domestic startup companies, create infant industries in the high-tech and IT sectors, and help create an investor class.

- Another important use of this capital is to address the demographic crunch. Countries in the region are facing a youth explosion that will put a strain on resources and the security apparatus. It is important that some of this extra capital is channeled toward job training, upgrading the educational systems, and improving "nationalization" of the job markets.

- The majority of businesses in the Gulf are family owned. With the demographic boom and the liberalization of economies, these businesses have to be able to compete at the global level. The economy and the livelihood of the citizen depend on these businesses, and a meaningful "transition" into public or private companies is needed for these businesses to survive and thrive in a global environment.

- Opening up the economy to trade is beneficial in the long-term, if it is managed responsibly. It also has its drawbacks. The same rules that allow for capital to flow in will allow it to flow out. Capital controls are hard to enforce and are often counterproductive, but countries in the region have to manage the flow of capital and prevent capital flight by building business-hospitable environments through limited regulations but strong enforceable laws.

- With the push to open up their economies, countries in the Gulf lack a clear comparative advantage in any sector other than energy. Saudi Arabia has announced plans to make the Kingdom the number one destination of foreign investment and create a vibrant financial sector. Other countries in the region, however, have opened up their economies for the sake of opening up with no clear long-term plans.

- The Gulf countries need to improve credibility and transparency of their monetary and fiscal policies, which is all the more important to attract foreign capital. International investors need to feel confident about the market and economic policies. The Saudi Monetary Agency has provided credible reports on an annual and a quarterly basis. However, there has been limited reporting on the part of the other GCC countries to provide the same level of transparency.

- Countries in the Gulf lack sound regulatory agencies. Hyping, dumping, and rumored investing are all too common in the Gulf. A GCC-wide agency or national regulatory bodies must be created in each country to monitor security, equity, and bond trading. In addition, the Gulf States must standardize their rules and regulations, especially if they hope to create a monetary union in 2010.

Even a casual observer must notice that recent growth in the Saudi and GCC stock markets is unprecedented and requires careful attention. It is important to note, however, that there have been some tangible efforts to improve the business environment in the Gulf States. For example, a recent report by the International Financial Cooperation (IFC) shows that these efforts have improved the competitiveness of some of the GCC economies. The IFC ranks Saudi Arabia as the most competitive country in the Arab world and the 38th globally. In one year, the Kingdom jumped 29 places among the 135 countries in contention. Kuwait followed as the 47th, Oman as the 51st, and the United Arab Emirates as the 69th.

However, the results of such theoretical ranking models are meaningless if they do not translate into real improvements on the ground. Dealing with the overvaluation of the stock markets and the risks outlined above is all too important to ensuring long-run economic stability.

The fact that the GCC states have implemented some meaningful market reforms does not mean these reforms will translate into realistic remedies to some serious risks in the various Gulf stock markets. One key area for improvement is to ensure that the GCC stock markets and their economies as a whole are strong enough to withstand speculative equity and currency attacks, and other issues are to channel the excess wealth into meaningful structural economic reforms and to deal with the most pressing issues, the demographic crunch, and the high unemployment rates.

THE DEMOGRAPHIC CRUNCH

Population growth is another major challenge to regional stability. Demographics compound the impact of low oil and gas export revenues on regional economies and increase the risk of political unrest. Oil income per capita drops because of the youth explosion discussed earlier. At the same time, some 40 percent of the region's population is now under 15 years of age, and rates of population growth are projected to be high enough in a number of countries so that the number of people entering the labor force will often double over a period of a decade. The region's educational system is already under extreme stress, and real and disguised unemployment for males between 18 and 25 years of age probably averages over 20 percent.[12]

These problems are further compounded by labor migration and the slow breakdown of the region's traditional family, clan, and tribal system, which is based on villages and the extended family. They are also compounded by hyperurbanization, a shift away from agriculture, and the need to absorb an increasingly well-educated population of women both for social reasons and to create productive economies that are globally competitive.

Virtually all southern Gulf States are heavily dependent on foreign labor at a time when many of their own younger citizens lack not only jobs but also the training and work ethic to get them. In many cases, these problems are reinforced by poor immigration policies that are routinely violated by the toleration of illegal immigrants, the issue of visas for money, and the existence of laws that require major benefit packages for native labor, thus making it difficult to hire or fire native labor. Some countries are trying to solve the problem with erratic purges of foreign labor, but most still lack consistent policies.

Unless average oil export revenues remain high in real terms over the coming decades, a combination of fluctuating oil revenues, high population growth rates, and a failure to modernize and diversify the overall economy could turn past oil wealth into oil poverty.

Until the recent oil boom, the southern Gulf States had only about 40 percent of the real per capita income they had at the peak of the oil boom in the early 1980s, and population growth played as important a role as poor economic policies and fiscal mismanagement. Kuwait, Qatar, and the United Arab Emirates did maintain high per capita incomes, but even Saudi Arabia's "oil wealth" is becoming increasingly marginal, as its population grows far more quickly than its economy.

Overall Trends in Population Growth

The MENA region has seen tremendous population growth during the last 50 years, and such growth is expected to continue. The total population of the Middle East and North Africa grew from 92.37 million in 1950 to 105.94 million in 1960, 138.91 million in 1970, 186.23 million in 1980, 254.17 million in 1990, 316.33 million in 2000, and 348.24 million in 2005.

This growth is expected to slow, but is still likely to have a massive future impact. Figure 3.14 shows the UN medium variant projection for 2004–2005. This is rather a conservative forecast and puts the MENA total population at 382.25 million in 2010, 452.22 million in 2020, 514.37 million in 2030, 568.44 million in 2040, and 611.96 million in 2050.

(In Millions)

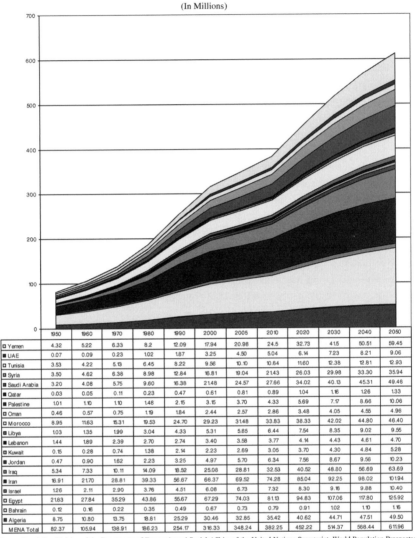

	1950	1960	1970	1980	1990	2000	2005	2010	2020	2030	2040	2050
Yemen	4.32	5.22	6.33	8.2	12.09	17.94	20.98	24.5	32.73	41.5	50.51	59.45
UAE	0.07	0.09	0.23	1.02	1.87	3.25	4.50	5.04	6.14	7.23	8.21	9.06
Tunisia	3.53	4.22	5.13	6.45	8.22	9.56	10.10	10.64	11.60	12.38	12.81	12.93
Syria	3.50	4.62	6.38	8.98	12.84	16.81	19.04	21.43	26.03	29.98	33.30	35.94
Saudi Arabia	3.20	4.08	5.75	9.60	16.38	21.48	24.57	27.66	34.02	40.13	45.31	49.46
Qatar	0.03	0.05	0.11	0.23	0.47	0.61	0.81	0.89	1.04	1.16	1.26	1.33
Palestine	1.01	1.10	1.10	1.48	2.15	3.15	3.70	4.33	5.69	7.17	8.66	10.06
Oman	0.46	0.57	0.75	1.19	1.84	2.44	2.57	2.86	3.48	4.05	4.55	4.96
Morocco	8.95	11.63	15.31	19.53	24.70	29.23	31.48	33.83	38.33	42.02	44.80	46.40
Libya	1.03	1.35	1.99	3.04	4.33	5.31	5.85	6.44	7.54	8.35	9.02	9.55
Lebanon	1.44	1.89	2.39	2.70	2.74	3.40	3.58	3.77	4.14	4.43	4.61	4.70
Kuwait	0.15	0.28	0.74	1.38	2.14	2.23	2.69	3.05	3.70	4.30	4.84	5.28
Jordan	0.47	0.90	1.62	2.23	3.25	4.97	5.70	6.34	7.56	8.67	9.56	10.23
Iraq	5.34	7.33	10.11	14.09	18.52	25.08	28.81	32.53	40.52	48.80	56.69	63.69
Iran	16.91	21.70	28.81	39.33	56.67	66.37	69.52	74.28	85.04	92.25	98.02	101.94
Israel	1.26	2.11	2.90	3.76	4.51	6.08	6.73	7.32	8.30	9.16	9.88	10.40
Egypt	21.83	27.84	35.29	43.86	55.67	67.29	74.03	81.13	94.83	107.06	117.80	125.92
Bahrain	0.12	0.16	0.22	0.35	0.49	0.67	0.73	0.79	0.91	1.02	1.10	1.16
Algeria	8.75	10.80	13.75	18.81	25.29	30.46	32.85	35.42	40.62	44.71	47.51	49.50
MENA Total	82.37	105.94	138.91	186.23	254.17	316.33	348.24	382.25	452.22	514.37	568.44	611.96

Source: Population Division of the Department of Economic and Social Affairs of the United Nations Secretariat, World Population Prospects: The 2004 Revision.

Figure 3.14   Total Population Trends in the MENA Region by Country: 1950–2050

Figure 3.15 shows the trend in the overall population growth by MENA country. The major countries, with large populations, are likely to see their population growth rates decline over the next 50 years. The UN estimates show the following:

• Algeria's annual percentage change in its total population has declined from 34 percent between 1980 and 1990 to 20 percent between 1990 and 2000, 8 percent between 2000

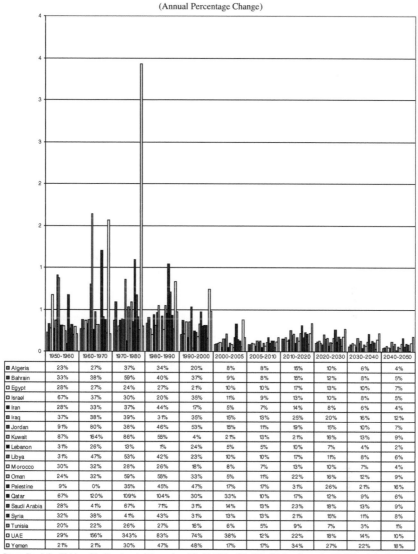

(Annual Percentage Change)

	1950-1960	1960-1970	1970-1980	1980-1990	1990-2000	2000-2005	2005-2010	2010-2020	2020-2030	2030-2040	2040-2050
⊟ Algeria	23%	27%	37%	34%	20%	8%	8%	15%	10%	6%	4%
■ Bahrain	33%	38%	59%	40%	37%	9%	8%	15%	12%	8%	5%
☐ Egypt	28%	27%	24%	27%	21%	10%	10%	17%	13%	10%	7%
☐ Israel	67%	37%	30%	20%	35%	11%	9%	13%	10%	8%	5%
■ Iran	28%	33%	37%	44%	17%	5%	7%	14%	8%	6%	4%
⊟ Iraq	37%	38%	39%	31%	35%	15%	13%	25%	20%	16%	12%
⊟ Jordan	91%	80%	38%	46%	53%	15%	11%	19%	15%	10%	7%
☐ Kuwait	87%	164%	86%	55%	4%	21%	13%	21%	16%	13%	9%
■ Lebanon	31%	26%	13%	1%	24%	5%	5%	10%	7%	4%	2%
⊟ Libya	31%	47%	53%	42%	23%	10%	10%	17%	11%	8%	6%
☐ Morocco	30%	32%	28%	26%	18%	8%	7%	13%	10%	7%	4%
☐ Oman	24%	32%	59%	55%	33%	5%	11%	22%	16%	12%	9%
■ Palestine	9%	0%	35%	45%	47%	17%	17%	31%	26%	21%	16%
■ Qatar	67%	120%	109%	104%	30%	33%	10%	17%	12%	9%	6%
■ Saudi Arabia	28%	41%	67%	71%	31%	14%	13%	23%	18%	13%	9%
■ Syria	32%	38%	41%	43%	31%	13%	13%	21%	15%	11%	8%
■ Tunisia	20%	22%	26%	27%	16%	6%	5%	9%	7%	3%	1%
☐ UAE	29%	156%	343%	83%	74%	38%	12%	22%	18%	14%	10%
☐ Yemen	21%	21%	30%	47%	48%	17%	17%	34%	27%	22%	18%

Source: Population Division of the Department of Economic and Social Affairs of the United Nations Secretariat, World Population Prospects: The 2004 Revision.

Figure 3.15   Population Growth Rates by Country Decline Over Time: 1950–2050

and 2005, 8 percent for 2005–2010, 15 percent for 2010–2020, 10 percent for 2020–2030, 6 percent for 2030–2040, and 4 percent for 2040–2050.

- Egypt's annual percentage change in its total population has declined from 27 percent between 1980 and 1990 to 21 percent between 1990 and 2000, 10 percent between 2000 and 2005, 10 percent for 2005–2010, 17 percent for 2010–2020, 13 percent for 2020–2030, 10 percent for 2030–2040, and 7 percent for 2040–2050.

- Iran's annual percentage change in its total population has declined from 44 percent between 1980 and 1990 to 17 percent between 1990 and 2000, 5 percent between 2000 and 2005, 7 percent for 2005–2010, 14 percent for 2010–2020, 8 percent for 2020–2030, 6 percent for 2030–2040, and 4 percent for 2040–2050.

- Iraq's annual percentage change in its total population has declined from 31 percent between 1980 and 1990 to 35 percent between 1990 and 2000, 15 percent between 2000 and 2005, 13 percent for 2005–2010, 25 percent for 2010–2020, 20 percent for 2020–2030, 16 percent for 2030–2040, and 12 percent for 2040–2050.

- Saudi Arabia's annual percentage change in its total population has declined from 71 percent between 1980 and 1990 to 31 percent between 1990 and 2000, 14 percent between 2000 and 2005, 13 percent for 2005–2010, 23 percent for 2010–2020, 18 percent for 2020–2030, 13 percent for 2030–2040, and 9 percent for 2040–2050.

The same trends are occurring in the smaller countries of the MENA region, and sometimes the change is even more striking in countries with a high influx of foreign workers such as the United Arab Emirates. The population growth rate for the United Arab Emirates between 1970 and 1980 was estimated to be 343 percent, by 2005 this rate is estimated to have declined to 38 percent, and by 2050 it will decline to 10 percent. These estimates are on the conservative side, but they are forecasts about an uncertain region of the world where the social, economic, and political dynamics are evolving.

The fact that a general decline in the growth rates is taking place also will not affect the fact the overall population is young and causing a youth explosion that will last at least two decades and that has many strategic, political, and economic implications far more serious than the general trends of the overall population growth rate.

The Youth Explosion

According to UN estimates made in 2005, the median ages for the MENA region are increasing but they remain high, as shown in Figure 3.16. It is estimated that they will increase from roughly 19.8 in 1950, to 18.8 in 1960, to 18.0 in 1970, to 18.9 in 1980, to 20.3 in 1990, to 22.8 in 2000, and to 24.1 in 2005.

According to the medium variant projections of the UN, the median age in the MENA region is estimated to be 25.6 in 2010, 28.7 in 2020, 31.8 in 2030, 34.7 in 2040, and 37.3 in 2050. If one compares these numbers to the rest of the world, one finds that even after a dramatic decline, the MENA region remains one of the youngest in the world. For example, compared to the MENA median age of 24.1 in 2005, the world has a median age of 28.1, the median age in Japan is 42.9, in Italy is 42.3, in Switzerland is 40.8, in the United States is 36.1, in France is 39.3, and in the United Kingdom is 39.0.[13]

This means that at least half of the population is under the age of 25. It is equally worth noting that this age group is increasing. The largest age group in the MENA population is under the age of 14. In 2005, the UN estimates that the percentage of the population that is under the age of 14 in the MENA countries ranged between

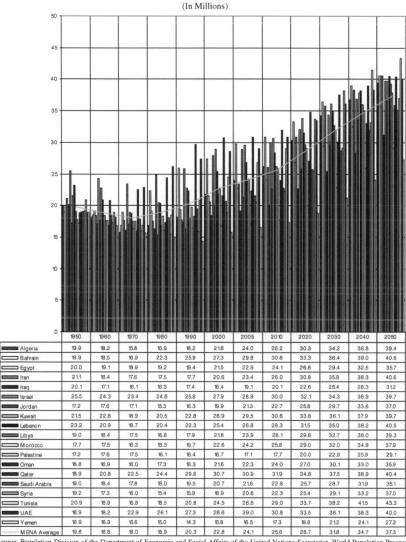

(In Millions)

	1950	1960	1970	1980	1990	2000	2005	2010	2020	2030	2040	2050
Algeria	19.9	18.2	15.8	16.9	18.2	21.8	24.0	26.2	30.3	34.2	36.8	39.4
Bahrain	18.9	18.5	16.9	22.3	25.9	27.3	29.8	30.8	33.3	36.4	39.0	40.6
Egypt	20.0	19.1	18.9	19.2	19.4	21.5	22.8	24.1	26.8	29.4	32.5	35.7
Iran	21.1	18.4	17.6	17.5	17.7	20.6	23.4	26.0	30.8	35.8	38.3	40.6
Iraq	20.1	17.1	16.1	16.3	17.4	18.4	19.1	20.1	22.6	25.4	28.3	31.2
Israel	25.5	24.3	23.4	24.8	25.8	27.9	28.9	30.0	32.1	34.3	36.9	39.7
Jordan	17.2	17.6	17.1	15.3	16.3	19.9	21.3	22.7	25.8	29.7	33.6	37.0
Kuwait	21.5	22.8	18.9	20.5	22.8	28.9	29.5	30.6	33.8	36.1	37.9	39.7
Lebanon	23.2	20.9	18.7	20.4	22.3	25.4	26.8	28.3	31.5	35.0	38.2	40.5
Libya	19.0	18.4	17.5	16.8	17.9	21.8	23.9	26.1	29.6	32.7	36.0	39.3
Morocco	17.7	17.5	16.3	18.3	19.7	22.6	24.2	25.8	29.0	32.0	34.9	37.9
Palestine	17.2	17.6	17.5	16.1	16.4	16.7	17.1	17.7	20.0	22.9	25.9	29.1
Oman	18.8	16.9	16.0	17.3	18.3	21.6	22.3	24.0	27.0	30.1	33.0	35.9
Qatar	18.9	20.8	22.5	24.4	29.6	30.7	30.9	31.9	34.8	37.5	38.9	40.4
Saudi Arabia	19.0	18.4	17.8	18.0	19.5	20.7	21.6	22.8	25.7	28.7	31.9	35.1
Syria	19.2	17.3	16.0	15.4	15.9	18.9	20.6	22.3	25.4	29.1	33.2	37.0
Tunisia	20.9	18.9	16.8	18.5	20.8	24.5	26.8	29.0	33.7	38.2	41.5	43.3
UAE	18.9	18.2	22.9	26.1	27.3	28.6	29.0	30.8	33.5	36.1	38.3	40.0
Yemen	18.9	16.9	15.6	15.0	14.3	15.8	16.5	17.3	18.8	21.2	24.1	27.2
MENA Average	19.8	18.8	18.0	18.9	20.3	22.8	24.1	25.6	28.7	31.8	34.7	37.3

Source: Population Division of the Department of Economic and Social Affairs of the United Nations Secretariat, World Population Prospects: The 2004 Revision.

Figure 3.16   Median Age Trends in the MENA Region by Country: 1950–2050

22 and 45 percent and projects that by 2050, this range will change to 14.5 and 26.3 percent.[14]

The following figures provide a graphic picture of age distribution in the MENA region:

- Figure 3.17 shows the UN estimates of each age group in the MENA region (0–85+) from 1950 to 2050. It does not only show the overall trend in the population growth that

(In Millions)

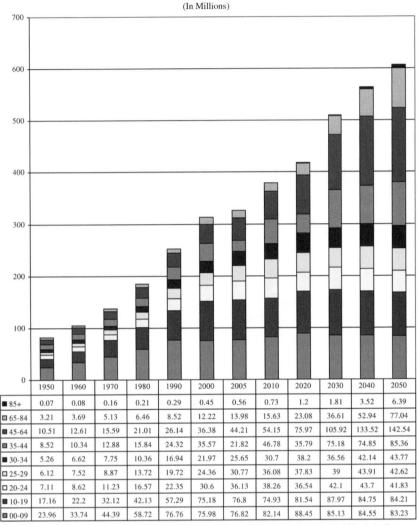

	1950	1960	1970	1980	1990	2000	2005	2010	2020	2030	2040	2050
■ 85+	0.07	0.08	0.16	0.21	0.29	0.45	0.56	0.73	1.2	1.81	3.52	6.39
▨ 65-84	3.21	3.69	5.13	6.46	8.52	12.22	13.98	15.63	23.08	36.61	52.94	77.04
■ 45-64	10.51	12.61	15.59	21.01	26.14	36.38	44.21	54.15	75.97	105.92	133.52	142.54
▨ 35-44	8.52	10.34	12.88	15.84	24.32	35.57	21.82	46.78	35.79	75.18	74.85	85.36
■ 30-34	5.26	6.62	7.75	10.36	16.94	21.97	25.65	30.7	38.2	36.56	42.14	43.77
▢ 25-29	6.12	7.52	8.87	13.72	19.72	24.36	30.77	36.08	37.83	39	43.91	42.62
▢ 20-24	7.11	8.62	11.23	16.57	22.35	30.6	36.13	38.26	36.54	42.1	43.7	41.83
■ 10-19	17.16	22.2	32.12	42.13	57.29	75.18	76.8	74.93	81.54	87.97	84.75	84.21
▨ 00-09	23.96	33.74	44.39	58.72	76.76	75.98	76.82	82.14	88.45	85.13	84.55	83.23

Source: Population Division of the Department of Economic and Social Affairs of the United Nations Secretariat, World Population Prospects: The 2004 Revision.

Figure 3.17 Population Distribution and the "Youth Explosion" in the MENA Region: 1950–2050

was discussed earlier, but it also shows the youth explosion in the region. The percentage of the 00–09 may decline, as a percentage of the total population, but the actual numbers do not. For example, in 1950, the MENA 00–09 population was 23.96 million, it increased to 76.76 million in 1990, 75.98 million in 2000, 76.82 million in 2005, and is projected to reach 82.14 million in 2010, 88.45 million in 2020, 85.13 million in 2030, 84.55 million in 2040, and 83.23 million in 2050.

Saudi Arabia   (In Thousands)

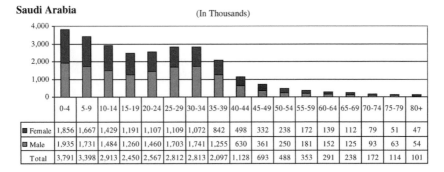

	0-4	5-9	10-14	15-19	20-24	25-29	30-34	35-39	40-44	45-49	50-54	55-59	60-64	65-69	70-74	75-79	80+
Female	1,856	1,667	1,429	1,191	1,107	1,109	1,072	842	498	332	238	172	139	112	79	51	47
Male	1,935	1,731	1,484	1,260	1,460	1,703	1,741	1,255	630	361	250	181	152	125	93	63	54
Total	3,791	3,398	2,913	2,450	2,567	2,812	2,813	2,097	1,128	693	488	353	291	238	172	114	101

Iran

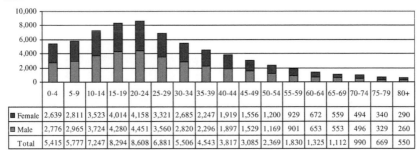

	0-4	5-9	10-14	15-19	20-24	25-29	30-34	35-39	40-44	45-49	50-54	55-59	60-64	65-69	70-74	75-79	80+
Female	2,639	2,811	3,523	4,014	4,158	3,321	2,685	2,247	1,919	1,556	1,200	929	672	559	494	340	290
Male	2,776	2,965	3,724	4,280	4,451	3,560	2,820	2,296	1,897	1,529	1,169	901	653	553	496	329	260
Total	5,415	5,777	7,247	8,294	8,608	6,881	5,506	4,543	3,817	3,085	2,369	1,830	1,325	1,112	990	669	550

Egypt

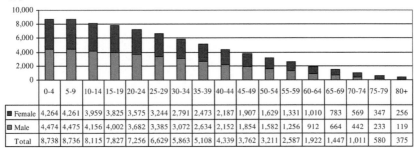

	0-4	5-9	10-14	15-19	20-24	25-29	30-34	35-39	40-44	45-49	50-54	55-59	60-64	65-69	70-74	75-79	80+
Female	4,264	4,261	3,959	3,825	3,575	3,244	2,791	2,473	2,187	1,907	1,629	1,331	1,010	783	569	347	256
Male	4,474	4,475	4,156	4,002	3,682	3,385	3,072	2,634	2,152	1,854	1,582	1,256	912	664	442	233	119
Total	8,738	8,736	8,115	7,827	7,256	6,629	5,863	5,108	4,339	3,762	3,211	2,587	1,922	1,447	1,011	580	375

Source: U.S. Census Bureau, International Data Base, April 2005 version

Figure 3.18   The Youth Explosion in Selected MENA Countries as Case Studies: 2005

• Figure 3.18 shows the U.S. Census Bureau estimates of three major MENA countries: Saudi Arabia, Iran, and Egypt in 2005. The overall patterns are the same. The population is largely young, and, in fact, the largest portion is that under the age of 4 in the case of Saudi Arabia. The same is true to a large extent in Egypt where the largest age population is between 5 and 9. Iran has a slightly different distribution, where the largest population group is between the age of 20 and 24. This does indicate that Iran may have a more urgent need to deal with its youth explosion.

This youth explosion will exhaust natural water supplies, force permanent dependence on food imports, and create an immense bow wave of future strain on the social, educational, political, and economic systems. Another equally important element in the MENA demographic dynamics is the overcrowding and the strain this population growth is putting on the physical infrastructure.

Figure 3.19 shows the rate of urbanization of the MENA region from 1960 until 2004. The UN estimates that 33.52 percent of the population lived in urban areas in

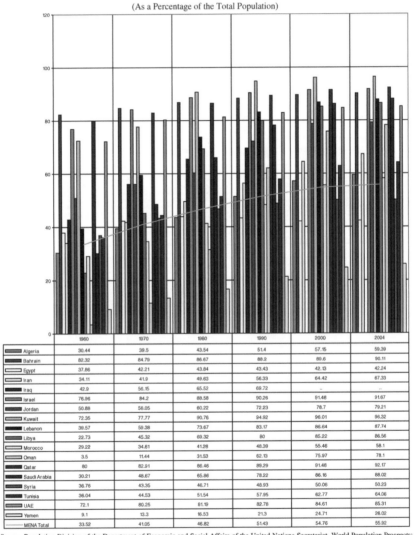

(As a Percentage of the Total Population)

	1960	1970	1980	1990	2000	2004
Algeria	30.44	39.5	43.54	51.4	57.15	59.39
Bahrain	82.32	84.79	86.67	88.2	89.6	90.11
Egypt	37.86	42.21	43.84	43.43	42.13	42.24
Iran	34.11	41.9	49.63	56.33	64.42	67.33
Iraq	42.9	56.15	65.52	69.72	..	..
Israel	76.96	84.2	88.58	90.26	91.48	91.67
Jordan	50.88	56.05	60.22	72.23	78.7	79.21
Kuwait	72.35	77.77	90.76	94.92	96.01	96.32
Lebanon	39.57	59.38	73.67	83.17	86.64	87.74
Libya	22.73	45.32	69.32	80	85.22	86.56
Morocco	29.22	34.61	41.28	48.39	55.46	58.1
Oman	3.5	11.44	31.53	62.13	75.97	78.1
Qatar	80	82.91	86.46	89.29	91.48	92.17
Saudi Arabia	30.21	48.67	65.86	78.22	86.16	88.02
Syria	36.76	43.35	46.71	48.93	50.06	50.23
Tunisia	36.04	44.53	51.54	57.95	62.77	64.06
UAE	72.1	80.25	81.19	82.78	84.61	85.31
Yemen	9.1	13.3	16.53	21.3	24.71	26.02
MENA Total	33.52	41.05	46.82	51.43	54.76	55.92

Source: Population Division of the Department of Economic and Social Affairs of the United Nations Secretariat, World Population Prospects: The 2004 Revision.

Figure 3.19   The Rate of Urbanization in the MENA Region by Country: 1960–2004

1960. This increased to 51.43 percent in 1990, to 54.76 percent in 2000, and to 55.92 percent in 2004. The reasons for this "hyperurbanization" are due to the pace of development and the demand for jobs, educational centers, and health-care services, and the overall decline in agriculture as a sustaining means to live. In addition to strains on the infrastructure, hyperurbanization and a half-century decline in agricultural and traditional trades imposed high levels of stress on traditional social safety nets and extended families—the basic social cohesion unit.

Population growth already presents major problems for future infrastructure development. Major problems now exist in every aspect of infrastructure from urban services to education to health care. At the same time, population pressure is exhausting natural water supplies in many countries, leading to growing dependence on desalination, and forcing permanent dependence on food imports. Demand for water already exceeds the supply in nearly half the countries in the region, and annual renewable water supplies per capita have fallen by 50 percent since 1960 and are projected to fall from 1,250 square meters today to 650 square meters in 2025—about 14 percent of today's global average. Groundwater is being overpumped and "fossil water" depleted.

The problems of the youth explosion and the trend toward urban centers are compounded by overstretched and outdated educational systems and the failure of the labor market to create productive jobs or any jobs at all for many of the young men entering the labor force.

Addressing the Unemployment Problem

Demographics have already made employment a critical issue, and the problem will grow steadily for at least the next quarter century because of existing population momentum. In addition, the youth explosion exacerbates the problem of unemployment. The implications of high unemployment go beyond their economic value; they can cause major instability problems and affect the perception of legitimacy of the governments in the MENA region.

The region again lags behind all other regions—including sub-Saharan Africa—in the level of employment, as is shown in Figure 3.20. According to the International Labor Organization, the unemployment rate in the MENA region was 12.1 percent in 1993, compared to 2.4 percent in East Asia, 4.8 percent in South Asia, 3.9 percent in Southeast Asia, 8.0 percent in the industrialized economies, 6.9 percent in Latin America and the Caribbean, 6.3 percent in transition economies, and 11.0 percent in sub-Saharan Africa. The MENA rate of unemployment increased to 12.25 percent in 2003 compared to 3.3 percent in East Asia, 4.8 percent for South Asia, 6.3 percent in Southeast Asia, 6.83 percent in industrialized economies, 8.0 percent in Latin America and the Caribbean, 9.2 percent in transition economies, and 10.9 percent in sub-Saharan Africa.[15]

These numbers may be conservative estimates because they do not take into account the "disguised" unemployment rates and the unproductive government jobs. Figure 3.21 provides rough estimates of the Gulf countries' direct and disguised

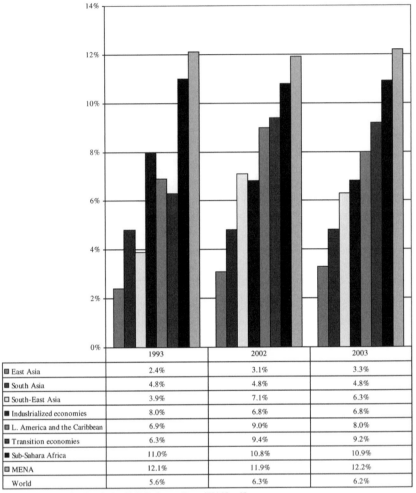

	1993	2002	2003
▣ East Asia	2.4%	3.1%	3.3%
■ South Asia	4.8%	4.8%	4.8%
☐ South-East Asia	3.9%	7.1%	6.3%
■ Industrialized economies	8.0%	6.8%	6.8%
▣ L. America and the Caribbean	6.9%	9.0%	8.0%
■ Transition economies	6.3%	9.4%	9.2%
■ Sub-Sahara Africa	11.0%	10.8%	10.9%
▣ MENA	12.1%	11.9%	12.2%
World	5.6%	6.3%	6.2%

Source: International Labor Organization, World Employment Report 2004-05, p. 27.

Figure 3.20   Global Unemployment Rates by Region: 1993–2003

unemployment. The disguised unemployment is sometimes as high as double that of the direct unemployment. These estimates are dated and were made in 2003. Nevertheless, they are still a warning about stability. They range from 15 to 45 percent, and the demographic trends mean that the majority of these rates are likely to continue for the foreseeable future. They differ in their growth rates, their population size, and the age distribution. They also differ in the size of their job market and in the number of jobs that have to be created every year to reduce their unemployment rates.

Figure 3.22 shows the CIA *World Fact Book 2005* estimates of the labor forces in the MENA countries. Iran leads the way with 23 million workers, Egypt with

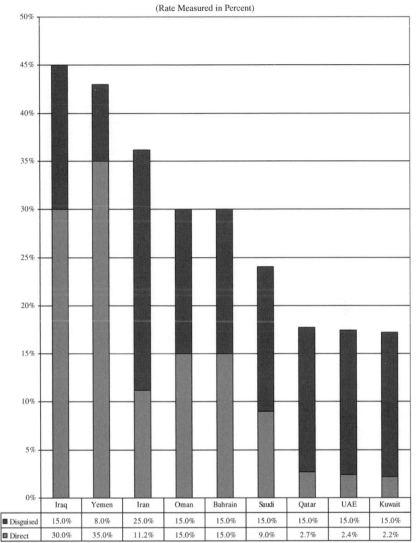

(Rate Measured in Percent)

	Iraq	Yemen	Iran	Oman	Bahrain	Saudi	Qatar	UAE	Kuwait
■ Disguised	15.0%	8.0%	25.0%	15.0%	15.0%	15.0%	15.0%	15.0%	15.0%
■ Direct	30.0%	35.0%	11.2%	15.0%	15.0%	9.0%	2.7%	2.4%	2.2%

Note: Disguised includes public sector, civil service, and private sector jobs with no use economic output.
Source: Rough estimates based on the CIA. World Fact Book, 2005.

Figure 3.21   A Rough Estimate of Direct and Disguised Unemployment Rates in the Gulf: 2005

20.7 million, Morocco with 11.02 million, Algeria with 9.9 million, Iraq with 6.7 million, Saudi Arabia with 6.6 million, Yemen with 5.98 million, Syria with 5.12 million, Israel with 2.68 million, Lebanon with 2.6 million, United Arab Emirates with 2.36 million, Libya with 1.59 million, Kuwait with 1.42 million, Jordan with 1.4 million, Oman with 0.9 million, Gaza and the West Bank with 1.8 million,

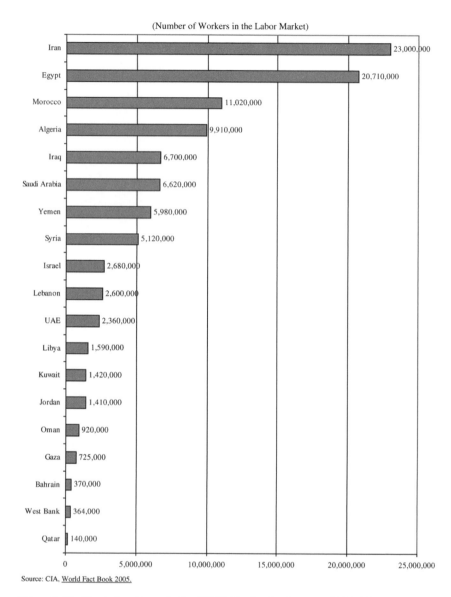

Figure 3.22   The Labor Force in the MENA Region by Country: 2004–2005

Bahrain with 0.370 million, and Qatar with 0.140 million. These are enormous numbers, especially for the large countries. This is compounded by the fact that their economies are not growing as rapidly to sustain reasonable employment levels and keep up with the youth explosion.

They also mean that extraordinary numbers of young workers are entering the job market. The CIA *World Fact Book* and the *Arab Human Development Report* both estimated that in 2002, the youth explosion was causing a massive influx of young males into the labor market. Figure 3.23 shows these estimates. For example, Iran had 801,000 young men entering the job market, Iraq had 260,000 young men entering the job market, and Saudi Arabia had 223,400 young men entering the job market.

These numbers also do not begin to tell the whole story because they do not take into account the number of women entering the job market, and global

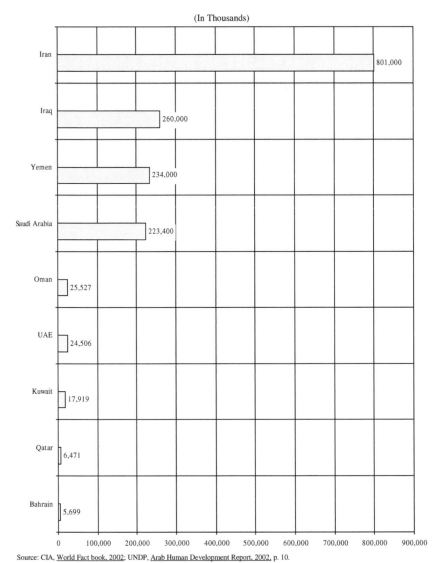

Source: CIA, World Fact book, 2002; UNDP, Arab Human Development Report, 2002, p. 10.

Figure 3.23   The Number of Young Males Entering the Labor Market Each Year

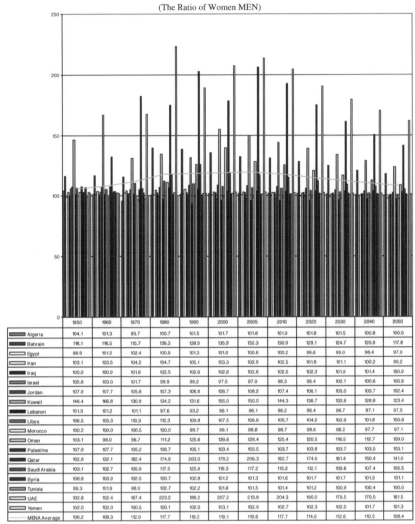

	1950	1960	1970	1980	1990	2000	2005	2010	2020	2030	2040	2050
Algeria	104.1	101.3	95.7	100.7	101.5	101.7	101.8	101.9	101.8	101.5	100.8	100.0
Bahrain	116.1	116.5	115.7	139.3	138.5	135.8	132.3	130.9	128.1	124.7	120.9	117.8
Egypt	98.9	101.2	102.4	100.9	101.3	101.0	100.6	100.2	99.8	99.0	98.4	97.9
Iran	103.1	103.5	104.2	104.7	105.1	103.3	102.9	102.5	101.8	101.1	100.2	99.2
Iraq	100.0	100.9	101.8	102.5	102.6	102.6	102.6	102.5	102.3	101.9	101.4	100.9
Israel	105.8	103.0	101.7	99.9	99.2	97.5	97.9	98.3	99.4	100.1	100.6	100.9
Jordan	107.9	107.7	105.6	107.3	108.8	108.7	108.2	107.4	106.1	105.0	103.7	102.4
Kuwait	146.4	166.8	130.9	134.2	131.6	155.0	150.0	144.3	138.7	133.9	128.8	123.4
Lebanon	101.3	101.2	101.1	97.6	93.2	96.1	96.1	96.2	96.4	96.7	97.1	97.5
Libya	106.5	105.3	110.3	112.3	109.8	107.5	106.6	105.7	104.2	102.9	101.8	100.8
Morocco	100.2	100.0	100.5	100.0	99.7	99.1	98.8	98.7	98.6	98.2	97.7	97.1
Oman	103.1	98.0	96.7	111.2	125.8	139.6	128.4	125.4	120.5	116.5	112.7	109.0
Palestine	107.9	107.7	105.2	106.7	105.1	103.4	103.5	103.7	103.8	103.7	103.5	103.1
Qatar	102.8	132.1	182.4	174.9	203.0	178.2	206.3	192.7	174.6	161.4	150.4	141.0
Saudi Arabia	103.1	102.7	105.9	117.5	125.9	119.3	117.2	115.2	112.1	109.6	107.4	105.5
Syria	106.9	103.9	102.5	100.7	100.8	101.2	101.3	101.6	101.7	101.7	101.5	101.1
Tunisia	99.3	101.9	98.0	102.7	102.2	101.8	101.5	101.4	101.2	100.9	100.4	100.0
UAE	102.8	102.4	167.4	223.2	189.2	207.2	213.8	204.3	190.0	179.5	170.5	161.5
Yemen	102.0	102.0	100.5	100.1	102.0	103.1	102.9	102.7	102.3	102.0	101.7	101.3
MENA Average	106.2	108.3	112.0	117.7	118.2	119.1	119.6	117.7	114.9	112.6	110.5	108.4

Source: Population Division of the Department of Economic and Social Affairs of the United Nations Secretariat, World Population Prospects:
The 2004 Revision.

**Figure 3.24   Gender Ratio: The Role of Women in the MENA Economies by Country:
1950–2050**

competitiveness requires the efficient use of women in the labor force. Figure 3.24
shows the UN estimates of the female/male ratio in the MENA region between
1950 and 2050. On average, women outnumber men by at least 6 percent, and it
is increasing. In 1950, the MENA average women/men ratio was 1.06, 1.18 in
1990, 1.19 in 2000, 1.19 in 2005, but is projected to decline to 1.08 in 2050. It also
means that the number of job applicants that was discussed earlier is double, and so
may be the unemployment rate if this is taken into account.

Women's participation in the job market has improved over the years in most MENA countries, but it is still low. Figure 3.25 shows the UN estimates of women's participation in each of the MENA countries. Despite the fact that women outnumbered men in nearly all the MENA countries, women represented only a small fraction of the labor forces. Figure 3.25 shows that women represented 21.3 percent of the MENA labor forces in 1960, 23.17 percent in 1970, 24.61 percent in 1980, 24.64 percent in 1990, 28.42 percent in 2000, and 30.21 percent in 2004.

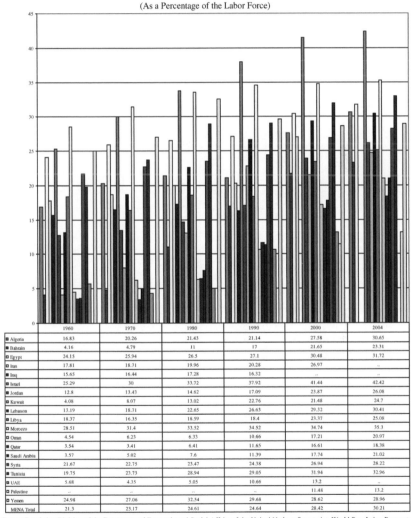

(As a Percentage of the Labor Force)

	1960	1970	1980	1990	2000	2004
Algeria	16.83	20.26	21.43	21.14	27.58	30.65
Bahrain	4.16	4.79	11	17	21.65	23.31
Egypt	24.15	25.94	26.5	27.1	30.48	31.72
Iran	17.81	18.71	19.96	20.28	26.97	..
Iraq	15.65	16.44	17.28	16.32	..	..
Israel	25.29	30	33.72	37.92	41.44	42.42
Jordan	12.8	13.43	14.62	17.09	23.87	26.08
Kuwait	4.08	8.07	13.02	22.76	21.48	24.7
Lebanon	13.19	18.71	22.65	26.63	29.32	30.41
Libya	18.37	16.35	18.59	18.4	23.37	25.08
Morocco	28.51	31.4	33.52	34.52	34.74	35.3
Oman	4.54	6.23	6.33	10.66	17.21	20.97
Qatar	3.54	3.41	6.41	11.65	16.61	18.38
Saudi Arabia	3.57	5.02	7.6	11.39	17.74	21.02
Syria	21.67	22.75	23.47	24.38	26.94	28.22
Tunisia	19.75	23.73	28.94	29.05	31.94	32.96
UAE	5.68	4.35	5.05	10.66	13.2	..
Palestine	..	..	..	..	11.48	13.2
Yemen	24.98	27.06	32.54	29.68	28.62	28.96
MENA Total	21.3	23.17	24.61	24.64	28.42	30.21

Source: Population Division of the Department of Economic and Social Affairs of the United Nations Secretariat, World Population Prospects: The 2004 Revision.

Figure 3.25   Women's Participation in the Labor Force by MENA Country: 1960–2004

More generally, immigration is another sign of current and future instability. The lack of job opportunities, dated educational systems, and the lagging economies have caused many in the MENA region to leave by immigration to other parts of the world. Figure 3.26 shows the estimated average rate of immigration in the MENA region between 1995 and 2050.

These estimates are based on the UN median variant projections in 2005. The trends are all too clear—people are leaving the MENA region in large numbers,

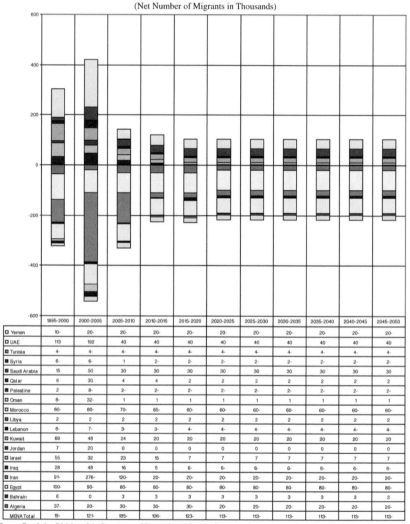

(Net Number of Migrants in Thousands)

	1995-2000	2000-2005	2005-2010	2010-2015	2015-2020	2020-2025	2025-2030	2030-2035	2035-2040	2040-2045	2045-2050
☐ Yemen	10-	20-	20-	20-	20-	20-	20-	20-	20-	20-	20-
☐ UAE	113	192	40	40	40	40	40	40	40	40	40
☐ Tunisia	4-	4-	4-	4-	4-	4-	4-	4-	4-	4-	4-
■ Syria	6-	6-	1	2-	2-	2-	2-	2-	2-	2-	2-
■ Saudi Arabia	15	50	30	30	30	30	30	30	30	30	30
■ Qatar	6	30	4	4	2	2	2	2	2	2	2
■ Palestine	2	8-	2-	2-	2-	2-	2-	2-	2-	2-	2-
☐ Oman	8-	32-	1	1	1	1	1	1	1	1	1
☐ Morocco	60-	80-	70-	65-	60-	60-	60-	60-	60-	60-	60-
☐ Libya	2	2	2	2	2	2	2	2	2	2	2
■ Lebanon	6-	7-	3-	3-	4-	4-	4-	4-	4-	4-	4-
☐ Kuwait	69	48	24	20	20	20	20	20	20	20	20
■ Jordan	7	20	0	0	0	0	0	0	0	0	0
☐ Israel	55	32	23	15	7	7	7	7	7	7	7
■ Iraq	28	48	16	5	6-	6-	6-	6-	6-	6-	6-
☐ Iran	91-	276-	120-	20-	20-	20-	20-	20-	20-	20-	20-
☐ Egypt	100-	90-	80-	80-	80-	80-	80-	80-	80-	80-	80-
■ Bahrain	6	0	3	3	3	3	3	3	3	3	3
☐ Algeria	37-	20-	30-	30-	30-	20-	20-	20-	20-	20-	20-
MENA Total	19-	121-	185-	106-	123-	113-	113-	113-	113-	113-	113-

Source: Population Division of the Department of Economic and Social Affairs of the United Nations Secretariat, World Population Prospects: The 2004 Revision.

Figure 3.26   Average Annual Net Number of Immigrants by Country: 1995–2050

and the number is growing. Between 1995 and 2000, there was a net loss of 19,000 who left the MENA region, this number jumped to 121,000 during 2000–2005, is estimated to reach 185,000 during 2005–2010, 106,000 during 2010–2015, 123,000 during 2015–2020, and 113,000 every five years between 2020 and 2050.

These numbers would be much larger if they were not offset by the large influx of foreign labor into the oil-producing states of the Gulf. For example, the UN estimates, for example, that between 1995 and 2000, there were 100,000 immigrants who left Egypt, but this number is compensated by 113,000 foreign workers who entered the United Arab Emirates.

This influx raises serious questions about the ability of the Gulf States to address their unemployment and security issues without making major reductions in the presence of foreign workers. The solution, however, has to be far more complicated than programs to force their hiring of native workers, exclude new foreign workers, and *de facto* deportation programs. The economies of labor importing countries are highly dependent on the skills, efficiency, and work ethic of foreign workers, and this creates a dilemma for the Gulf States in choosing between native labor and productive labor.

The Foreign Labor Dilemma in the Gulf States

The Gulf States have promised for more than a decade to deal with their high dependence on foreign labor. Ironically, some felt that other MENA security problems might help. Some people anticipated that the Gulf was becoming less hospitable to the foreign workers and that they would leave to safer parts of the world. This has not happened—partly because the terrorism threat has minimized in the last couple of years and partly because there is no other region in the world that can absorb that many workers. Bahraini Minister of Labor Majeed bin Mohsen Al-Alawi put the number of foreign workers in the Gulf countries at 12 million, "which is increasing by five percent every year and expected to reach 18 million in 12 years."[16]

The larger question, however, is just how dependent the Gulf States are on foreign workers. There are uncertainties in the available estimates. Figure 3.27, for example, shows the UN estimates of foreign population in the GCC states. It is clear that a number of GCC countries retain large numbers of foreign workers as a percentage of the total population. Qatar leads the way with 70 percent, 68 percent in the United Arab Emirates, 49 percent in Kuwait, 38 percent in Bahrain, 26 percent in Oman, and 24 percent in Saudi Arabia.

Other sources are different. Figure 3.28 shows the estimates of four respected sources on the portion of foreign population in the GCC states. The general ranking is consistent: United Arab Emirates, Qatar, Kuwait, Bahrain, Oman, and Saudi Arabia. The United Arab Emirates has 63 to 81 percent, Qatar has 60 to 75 percent, Kuwait has 55 to 65 percent, Bahrain has 34 to 40 percent, Oman has 19 to 27 percent, and Saudi Arabia has 21 to 30 percent.

The same is true for data on the origin of the foreign workers. Figure 3.29 shows the estimates done by Andrzej Kapiszewski in 2003, and they show the total number

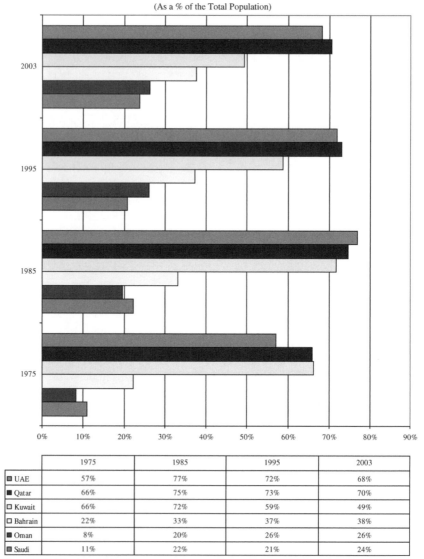

(As a % of the Total Population)

	1975	1985	1995	2003
▣ UAE	57%	77%	72%	68%
■ Qatar	66%	75%	73%	70%
□ Kuwait	66%	72%	59%	49%
□ Bahrain	22%	33%	37%	38%
■ Oman	8%	20%	26%	26%
▣ Saudi	11%	22%	21%	24%

Source: UN, World Population Policies 2003.

Figure 3.27   UN Estimates of Foreign Population Trends in the GCC Countries: 1975–2003

of foreign workers by their origin. There are, for example, 1.4 million and 1.0 million Indians in Saudi Arabia and the United Arab Emirates, respectively. There are also 250,000 Indonesians and 250,000 Sudanese in Saudi Arabia, which are mainly servants and drivers. These numbers may never go down and are unlikely to be replaced by Saudis.[17]

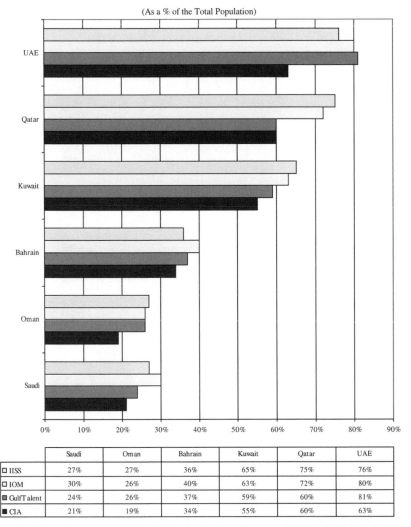

(As a % of the Total Population)

	Saudi	Oman	Bahrain	Kuwait	Qatar	UAE
☐ IISS	27%	27%	36%	65%	75%	76%
☐ IOM	30%	26%	40%	63%	72%	80%
▩ GulfTalent	24%	26%	37%	59%	60%	81%
■ CIA	21%	19%	34%	55%	60%	63%

Source: GulfTalent.com, Gulf Compensation Trends 2005, September 2005; IISS, Military Balance, 2005-2006; CIA, World Fact Book 2005; and Andrzej Kapiszweski, "Arab Labor Migration to the GCC States," IOM, September 2003.

Figure 3.28   Foreign Population in the GCC Countries by Source: 2004–2005

Figure 3.30 shows the International Institute for Strategic Studies (IISS) estimates of the origin of foreign workers. The IISS estimates that 60 percent of the United Arab Emirates is from Asia (the majority are from the Indian subcontinent), 36 percent of Qatar are Asian, 20 percent of Saudi Arabia are Asian, and 9 percent of Kuwait are Asian. Arabs from other countries also represent a large portion of the GCC population: 12 percent of the United Arab Emirates, 35 percent of Kuwait, 10 percent of Bahrain, and 6 percent of Saudi Arabia. The IISS does not provide a

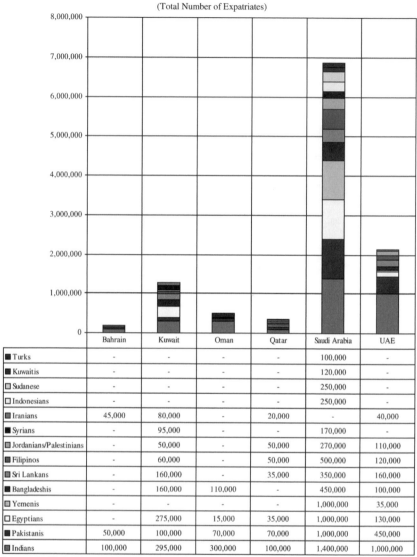

(Total Number of Expatriates)

	Bahrain	Kuwait	Oman	Qatar	Saudi Arabia	UAE
■ Turks	-	-	-	-	100,000	-
■ Kuwaitis	-	-	-	-	120,000	-
□ Sudanese	-	-	-	-	250,000	-
□ Indonesians	-	-	-	-	250,000	-
■ Iranians	45,000	80,000	-	20,000	-	40,000
■ Syrians	-	95,000	-	-	170,000	-
▨ Jordanians/Palestinians	-	50,000	-	50,000	270,000	110,000
▨ Filipinos	-	60,000	-	50,000	500,000	120,000
▨ Sri Lankans	-	160,000	-	35,000	350,000	160,000
■ Bangladeshis	-	160,000	110,000	-	450,000	100,000
□ Yemenis	-	-	-	-	1,000,000	35,000
□ Egyptians	-	275,000	15,000	35,000	1,000,000	130,000
■ Pakistanis	50,000	100,000	70,000	70,000	1,000,000	450,000
▨ Indians	100,000	295,000	300,000	100,000	1,400,000	1,000,000

Source: Andrzej Kapiszweski, "Arab Labor Migration to the GCC States," IOM, September 2003, p. 14.

Figure 3.29 Foreign Population in the GCC Countries by Country of Origin: 2002

breakdown of the Oman foreign population, but it is likely to mirror Saudi Arabia. In addition, it does have one of the lowest numbers of foreign labor workers, as a percentage of its total population.

What all of these estimates do have in common is that they show that past efforts to diversify the GCC dependence on foreign labor have not been as successful as hoped, although some countries have been more successful than others. Oman and

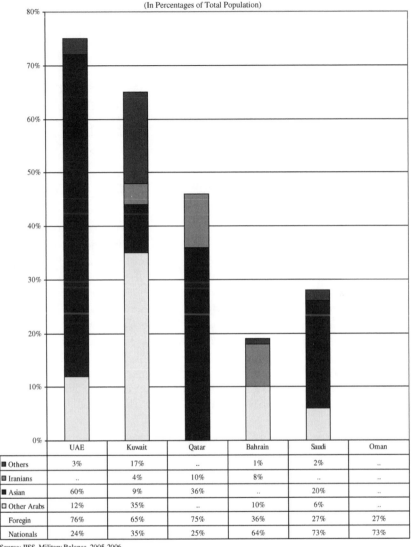

	UAE	Kuwait	Qatar	Bahrain	Saudi	Oman
■ Others	3%	17%	..	1%	2%	..
▣ Iranians	..	4%	10%	8%	..	..
■ Asian	60%	9%	36%	..	20%	..
□ Other Arabs	12%	35%	..	10%	6%	..
Foregin	76%	65%	75%	36%	27%	27%
Nationals	24%	35%	25%	64%	73%	73%

Source: IISS, Military Balance, 2005-2006.

Figure 3.30   Foreign Population Breakdown in Selected Gulf Countries: 2005–2006

Saudi Arabia have had done effective jobs of decreasing their dependence, but the United Arab Emirates and Qatar have increased their dependence over the last decade. As mentioned earlier, the population of the United Arab Emirates actually tripled between 1970 and 1980, and Qatar's doubled. This was largely due to the influx of foreign labor.

In addition to the economic and social pressures the presence of foreign workers is exerting on the GCC states, they can be a threat to domestic security. In recent years, many in the GCC states have attempted to explain the danger of having such a large foreign population among them. Officials in the GCC have also started to discuss ways to increase the effectiveness of their nationalization programs.

This helps explain why recent GCC summits have made the demographic issue front and center. The Saudi Labor Minister, Ghazi Al-Gosaibi, also warned that at one point there may be international pressure to "naturalize" the foreign population in the GCC. He was quoted as saying, "We have made the proposal fixing six years as the maximum period of stay for expatriates in order to deal with the demographic danger and the possibility of an international decision calling for naturalization of expatriates who have lived more than five years in the member states."[18]

THE IMPACT OF "DOMESTIC UNCERTAINTY" ON STRATEGIC STABILITY

At present, many MENA oil-exporting states can get by in spite of their problems. If low or low-to-moderate oil revenues should suddenly become the normal long-term case again, however, the resulting cut in government revenues will force many such countries to cut their budgets and development plans in ways that result in significant economic, social, and political trade-offs.

The International Monetary Fund stated in May 1998 that the decline in oil export revenues "would pose a serious risk to the growth outlook" for the Gulf region, "and particularly for the region's largest oil exporters such as Saudi Arabia and Kuwait...if sustained." This warning is just as true today, and it is clear from the economic history of the region, and virtually every current projection of future oil revenues, that population growth will outpace increases in oil revenues and cut per capita oil wealth indefinitely into the future.

The Impact of Oil Revenues on Internal Stability

If oil revenues are a blessing in terms of past and present income and development, the previous charts have shown they can also be a curse if nations rely on oil rather than on diversified development. This could be even truer in the future. In spite of the current and projected rises in MENA energy exports, the resulting export revenues will not meet the needs of Middle Eastern states with high population growth and economies with limited diversification unless average prices remain high, and even then, such revenues may fall far short of the need in many energy exporting states.

Violent swings between "oil crash" and "oil boom" have not helped the situation. They tend to undercut economic reform in "boom" years and make it unaffordable or politically impossible in bust years. Figure 3.31 presents a graphical representation of the busts and booms of the MENA oil revenues in constant 2000 dollars between 1971 and 2006.

One such "oil bust" took place in 1997–1999 only to be followed by a short oil boom, which was followed by relatively high prices. The oil crash that began in

(In Constant $US 2000 Billions)

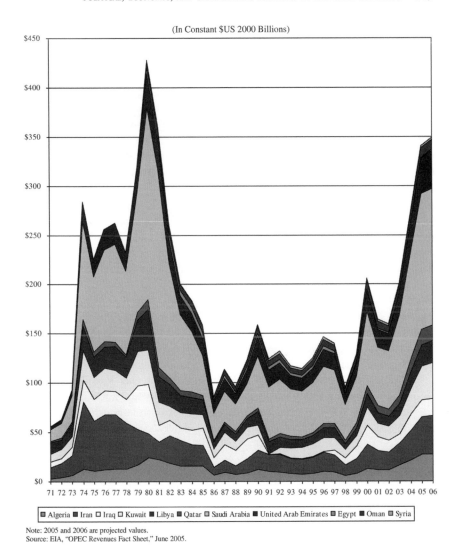

Note: 2005 and 2006 are projected values.
Source: EIA, "OPEC Revenues Fact Sheet," June 2005.

Figure 3.31   EIA Estimates of MENA Oil Revenues: Trends in Constant Dollars

1997 led to a series of unexpected cuts in oil prices that reached lows of $10 a barrel and cuts in annual oil revenues that approached 30 to 40 percent. As a result, the Organization of the Petroleum Exporting Countries (OPEC) oil revenues swung from $148.7 billion in 1997 to $99.9 billion in 1998. These cuts in oil revenues affected every major oil and gas producer in the Middle East and have reduced the region's ability to maintain both welfare payments and entitlements, and short-term investment.

The oil crash of 1997–1998 had a particularly dire impact on those MENA economies that had failed to modernize and diversify and/or were affected by the impact of sanctions on several critical suppliers. They led to sharp cuts in the estimated size of the future demand for exports, in national policies to increase production and export capacity, and the ability to obtain the investment necessary to implement those policies. They affected political stability and influenced a wide range of social problems, most importantly the impact of very high rates of population growth, the inability to sustain past welfare and entitlement programs, and the need to create new economic structures that offer suitable employment and incentives for investment.

In 1998, the OPEC oil revenues were as low as $123.2 billion in 2005 constant dollars. The cycle soon swung back in the other direction. Upward swings in oil revenues began to ease the situation in the spring of 1999. In March 1999, OPEC's member countries, together with some important outside producers, settled on a program of stringent oil production cuts. Following the implementation of cutbacks, the price of crude oil rose sharply over the course of 1999 and eventually reached levels in 2000 that had not been seen since the 1990–1991 Gulf War. They then dropped back to $190.7 billion in 2001 and $187 billion in 2002, with an estimated total of $223 billion in 2003.[19]

These swings in oil revenues are typical of other past cycles in oil revenues that contributed to the problems in Middle Eastern economic growth and the problems the region faces in dealing with its youth explosion and in funding both future development and expanded petroleum production and exports.

OPEC oil revenues were worth around $102.8 billion in constant 2000 dollars in 1972. After the October War and the 1974 oil embargo, they leaped to levels of around $443.4 billion and then dropped back to an average of $365.5 billion during 1975–1978. The fall of the Shah of Iran and the start of the Iran-Iraq War drove them to a new peak in 1980, when they were worth $597.5 billion. An oil price collapse began in 1985, and revenues dropped to $117.2 billion in 1986. They gradually rose back to levels of around $171.46 billion a year in early 1997, but a new oil crash began late that year. Major production cuts led to a rise in oil prices in 1999, but total revenues in nominal dollars still reached only $143.9 billion in 1999 and $226.6 billion in 2000. They were $117.2 billion in $U.S. 2000 in 2002.[20] OPEC oil revenues were estimated to be $429 billion in 2005 and $439.1 billion in 2006 (in 2005 constant dollars).[21]

Many countries are beginning to rethink their plans to increase production capacity and their attitudes toward private and foreign investment, but these swings in revenue lead to caution and plans that generally fall short of the kind of demand-drive supply projections made by the IEA and the EIA.

The region's economic and budget problems, and its energy investment capacity, have also been shaped by years of overreliance on oil wealth, economic mismanagement, massive population growth, mismanaged government spending, and the failure of regional governments to realistically plan and budget for the future.

Some key Middle Eastern governments have had a decade of nearly continuous budget deficits. Saudi Arabia and Iraq are key cases in point. Other countries are in major structural crisis. They cannot afford to implement their five-year plans and cannot fund both their present levels of entitlements and investment. Cases in point include Algeria, Syria, Bahrain, Iran, Oman, and Yemen. Most Middle Eastern governments now face a major short-term budget crisis, and this seems to include even states with relatively high ratios of exports to population: Kuwait, Qatar, and the Emirates other than Abu Dhabi and possibly Dubai.

Past drops in oil revenue and the resulting budget problems have already led to underinvestment in infrastructure, economic diversification, and state industries other than the petroleum sector in many states. Even the petroleum sector has been underfunded in some cases, although "starving the hand that feeds you" presents obvious enough problems for most Middle Eastern states to think twice.

Nevertheless, recent developments in the global energy market may change this situation strikingly in the future if MENA states experience *sustained* high average energy export revenues. This inflow of capital will also be met by a high propensity to invest by the citizens of the Middle East, allowing the MENA countries to maintain moderate levels of domestic investment. The question that still remains even if energy export revenues are high, however, is, will the MENA countries repeat the mistakes of the 1980s and 1990s of mismanaging oil revenues, or will they use their newfound "oil wealth" to implement realistic structural economic reforms? Furthermore, reforming welfare systems of entitlements, solving the unemployment problems, and reducing the Gulf States' reliance on foreign labor will be as important to stability as any political or security reforms.

Ability to Fund Investments to Increase Oil and Gas Production

As is discussed in more detail in Chapter 1 as well as in later chapters, the uncertainties surrounding future demand and future oil and gas export revenues do more than affect regional stability in ways that could lead to oil interruptions. They may also be creating serious long-term problems in financing the expansion of MENA oil and gas production capacity.

Unfortunately, an examination of current estimates of energy investment costs indicates that there is insufficient effort to realistically estimate the cost of the required future regional and country-specific investment requirements beyond relatively near-term projects to determine how well countries can finance development on their own through cash flow, loans, and various cash back or production-sharing arrangements.

Two critical factors could affect the ability to fund investment in increased oil and gas production. One is the growing limits on the budgets and investment capabilities of MENA energy exporting states caused by the lack of economic diversification and moderate oil prices. The second is the slowly increasing structural economic problem caused by rising populations, high welfare and entitlement programs, high military and arms expenditures, and low long-term revenues.

Market forces and state-driven energy investment may still be enough. Most Middle Eastern states have been relatively successful in using state revenues to fund energy investments in the past. It now seems likely, however, that their cash flow and savings will not be adequate to meet both their other spending and investment needs and energy investment needs.

The era of being able to safely rely on state oil and gas revenues to fund other state expenditures and investments may be over. Middle Eastern governments do not need to abandon state industries, state investment, and state control over energy resources, but fundamental reforms are needed to increase the ratio of foreign and domestic private investments. There currently, however, is no Middle Eastern country where market forces are allowed to operate without serious state interference. However, only a few oil-exporting countries—Bahrain, Egypt, Qatar, and Oman—are making serious progress. This helps explain why nearly all oil-producing countries in the Middle East are currently examining ways in which to privatize some aspects of their energy investment and obtain foreign investment.

At this point in time, there is no meaningful way to predict whether Middle Eastern oil-exporting states will persist in these plans if oil and gas revenues rise, how successful they will be in obtaining the energy investment and other capital they need, how much money any given country requires, and how well investments will be managed. Middle Eastern regimes tend to back-pedal on reform the moment oil revenues decline to moderate levels, and many face resistance from nationalists, Pan-Arab socialists, state-oriented technocrats, and Islamists. Virtually all states want to maximize revenues, but also have powerful elements that want to conserve resources for the future.

In short, foreign investment and the domestic private sector may have to assume a much larger share of the burden than many MENA countries now plan if the region is to produce the energy output the world needs. Relying on market forces might still lead to enough cost-effective investment, particularly given the oil industry's history of investing in reserves, future market share, and development even in periods of low oil income.